CAD/CAE Descriptive Geometry

Daniel L. Ryan
Clemson University, South Carolina

CRC Press
Boca Raton Ann Arbor London

Acquiring Editor: Russ Hall
Production Director: Sandy Pearlman
Production Services: Cracom Corporation
Cover Design: Miriam Recio Miller

Library of Congress Cataloging-in-Publication Data

Ryan, Daniel L., 1941–
CAD/CAE descriptive geometry/Daniel L. Ryan.
p. cm.
Includes index.
ISBN 0-8493-4273-2
1. Computer-aided engineering. 2. Geometry, Descriptive—Data processing. 3. Computer-aided design. I. Title
TA345.R89 1992 91-26323
604.2′01′5166—dc20 CIP

This book represents information obtained from authentic and highly regarded sources. Reprinted information is quoted with permission, and sources are indicated. Every reasonable effort has been made to give reliable data and information, but the authors and the publisher cannot assume responsibility for the validity of all materials or for the consequences of their use.

Direct all inquiries to CRC Press, 2000 Corporate Blvd., N.W., Boca Raton, Florida, 33431.

International Standard Book Number 0-8493-4273-2

Library of Congress Card Number 91-26323

Printed in the United States of America 1 2 3 4 5 6 7 8 9 0

Contents

Preface

Modern descriptive geometry is a mixture of plane geometry (mathematics), orthographic projection (technical drawing), and high-speed communication methods (digital computing). For readers who are unfamiliar with parts of this mixture, this book will introduce the fundamentals of each and the application of these processes. For those readers with some background, this book will appear slow paced and very elementary; the topics covered in this book are for those who want a sound foundation in fundamentals.

Constructive geometry, as we use it today, evolved from the transition between the industrial and the electronic revolutions. The process began before the industrial revolution and followed progress into the present electronic age. Therefore, constructive geometry is part of a design language and is used for one or more of the following purposes: (1) study, (2) construction, (3) maintenance, or (4) creation of an industrial product or project. When constructive geometry is used inside a digital computer, it becomes part of computer-aided design *(CAD)*. We use CAD to produce industrial drawings of all types and descriptions. In order to produce these drawings, we must understand where modern descriptive geometry came from, what it is now, and where it might lead us in the future.

Constructive geometry within CAD is commonplace in American industry today. Because CAD is taken for granted in the industrial marketplace and used whenever necessary, it is often overlooked in the college curriculum. The most visual part of CAD is computer graphics, which is covered in most curricula. No attempt has been made here to present computer graphics as a subject; computer graphics are used throughout the book as the tool for presentation of the constructive geometry used in CAD.

The selection of material in this textbook was based on the premise that the reader has access to a personal computer or PS/2 compatible CAD workstation running Autodesk, Inc., software. Therefore, each chapter presents a basic geometry topic and covers it in detail using the CAD workstation. It is the author's belief that this combination is necessary in an engineering or technology course because CAD requires ingenuity, inventiveness, imagination, and patience. An ability to use constructive geometry in design solutions can be practiced after exposure to the chapter materials.

CAD/CAE Descriptive Geometry

1 Descriptive Geometry Conventions

Modern descriptive geometry is a mixture of plane geometry (mathematics), third quadrant orthographic projection (engineering drawing), and high-speed communication methods (digital computing). For readers who are unfamiliar with parts of this mixture, this book will introduce the fundamentals of each and the application of these constructive geometry processes. To those readers with some background, this book will appear slow paced. The topics covered in this book are for those who want a sound foundation in fundamentals.

Constructive geometry, as we use it today, evolved from the transition into the electronic revolution. This chapter begins before the electronic revolution and follows the progress into the present electronic age. Therefore, constructive geometry is part of a design language and is used for one or more of the several purposes of (1) study, (2) construction, (3) maintenance, or (4) creation of an industrial product or project (computer-aided engineering [CAE]). When descriptive geometry is used inside a digital computer, it becomes part of computer-aided design (CAD). We use CAD to produce industrial drawings of all types and descriptions. In order to produce these drawings we must understand where modern descriptive geometry came from, what it is now, and where it might lead us in the future.

1.1 DESCRIPTIVE GEOMETRY HISTORY

One of the earliest evidences of descriptive geometry is found in the book on architecture by Marcus Vitruvius, who lived at the time of Augustus and was appointed superintendent of military designs for the Roman army. In his book, Marcus Vitruvius dealt with the concept of horizontal and vertical planes of construction of public buildings. This was the first recorded orthographic projection concept.

The first known mathematical principle of descriptive geometry appears in Albrecht Dürer's book *Geometry and Perspective,* published in Nuremberg in 1525. The first recorded three-view, first quadrant drawing appears in this work. It would be 200 years (1738) before another army officer Francois Frezier would publish a treatise on stereotomy for stonecutting. But this almost forgotten treatise was read by a brilliant young French mathematician Gaspard Monge. Monge was employed at the military school at Mezieres in 1768. He designed the fortifications used by the French army. While working on the many computations required, he discovered a geometrical solution for defilement. The ordinary mathematical procedures for the same solutions were very laborious. When Monge first presented his procedures in which he associated precise and mathematical elements to several common-scale elevation and plan drawings, the French army was skeptical of these simple and direct methods. After testing, his methods became a French military secret, and Monge was prohibited from revealing them until he lectured at the Ecole Polytechnique in 1795. His book on descriptive geometry was published in 1801.

One of Monge's students at the Polytechnique, Claude Crozet, was employed in 1816 to be an instructor of descriptive geometry at West Point. In 1821 the first United States-based information on descriptive geometry was published, and in 1826 Charles Davies, who succeeded Crozet, published the first text book on descriptive geometry. Modernizations of this first text were published in 1860 and 1865 by Professors Warren and Church and were used until 1910 when Henry Miller of the University of Michigan introduced a series of editions lasting until 1945. Miller's work was the last textbook effort to preserve the Monge abstract principles. The texts appearing after 1945 introduced more direct methods for solving abstract geometrical relationships. Noteworthy among the authors of such texts were Street, Grant, Pare, Loving, Hill, Hood, Palerlee, Springer, Hoelscher, plus a score of present-day authors.

1.2 DEFINITIONS

The process of learning modern descriptive geometry can be mastered in three steps:

1. *Comprehension:* The subject matter is in the form of printed words, diagrams, observations of demonstrated skills, and CAD models.
2. *Clarification:* The learner carefully reads and considers step by step what is perceived through the comprehension/presentation (step 1).
3. *Retention:* Further study, the solution of practice problems, and participation in evaluation methods determine the amount of clarification (step 2) that has taken place.

In the above three steps there will be a need for a system of written and graphical symbols. These symbols appear throughout this textbook and will appear at the CAD workstation used for the practice problems. It is important, therefore, that each reader make this chapter a reference dictionary to be used throughout the remainder of the text. Do not attempt to memorize the material presented in this chapter; instead you should refer back to it as needed.

1.3 ABBREVIATIONS

In a short time all abbreviations will have been memorized from constant use; for now refer back to them as often as needed.

Written Symbol	Graphical Symbol
AP = apex	Λ
AU = auxiliary	α
AX = axis (major: AX, minor: ax)	A
BA = base	B
BL = brilliant line	β
BP = brilliant point	γ
CN = cone	C

Written Symbol			Graphical Symbol
CV	=	convolute	ψ
CP	=	coordinate plane	ρ
CO	=	coplanar	l
CY	=	cylinder	o
DG	=	diagonals	/
DL	=	direct light	L
EL	=	element	E
EV	=	edge view	σ
FP	=	frontal plane (reference frontal: F)	F
GX	=	generatrix	G
GL	=	ground line	-
HP	=	horizontal plane (reference horizontal: H)	P
IL	=	indirect light	I
IN	=	intersection	X
LS	=	line of shadow	S
NL	=	normal	N
OB	=	oblique	◇
OP	=	orthographic projection	=
PA	=	parallel (parallelism)	//
PD	=	perpendicular (perpendicularity)	⊥
PL	=	plane	■
PP	=	profile plane (reference profile: P)	□
PR	=	projection	●
PT	=	point	⇩
RY	=	rays	R
RT	=	right	r
RV	=	revolution	θ
SF	=	surface	Σ
SH	=	shade	δ
SP	=	spheroid	T
SW	=	shadow	Θ
TA	=	tangent	τ
TR	=	trace	Ω
VP	=	vanishing point	x
VC	=	visual cone	Ψ
VR	=	visual ray	ζ

1.4 NOTATIONS

You will notice that most of the figures, diagrams, and monitor displays include the listed two-letter abbreviations, graphical symbol, and a conventional system of line segments, which are

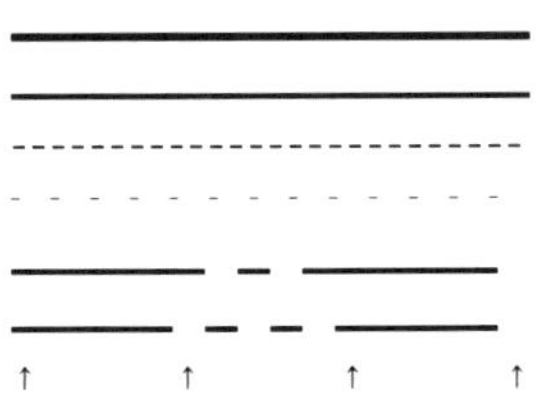

Object
Construction lines
Projection lines
Hidden lines
Center lines, traces, rotation lines
Axes of revolution, reference lines
Direction of sight

1.5 SYSTEM OF MEASUREMENT

Three mutually perpendicular planes, one horizontal (*HP*) and two vertical (*FP* and *PP*) are shown in Figure 1.1. These reference planes serve as the system of measurement or coordinate planes for the description and storage of points, lines, planes, and solids. These reference planes (*HP*, *FP*, and *PP*) intersect in the three axes labeled 0*X*, 0*Y*, and 0*Z* in Figure 1.1. The origin (0) is the intersection of the three axes. Four quadrants are formed by the intersecting axes, namely

1—where *X* and *Y* measurements are both positive (+),
2—where *X* is negative (−) and *Y* is positive,
3—where *X* is negative and *Y* is negative and,
4—where *X* is positive and *Y* is negative.

Measurements are made to points in one of these four quadrants where

X = distance to point right or left of 0, as $+X$ or $-X$,
Y = distance to point above or below 0 as $+Y$ or $-Y$ and,
Z = distance to point before or behind 0, as $+Z$ or $-Z$.

The point measurements (coordinates) are always given in *X*, *Y*, and *Z* order, and the distances are displayed in graphic display units (GDU). A GDU is set at the CAD workstation and can be anything from SI (metric) units to English feet and inches for architectural applications. For this book, the units of measurement are all in millimeters, unless otherwise noted. So, a point located (25,50,75) is 1 inch (25 millimeters) to the right, 2 inches above, and 3 inches in front of the origin.

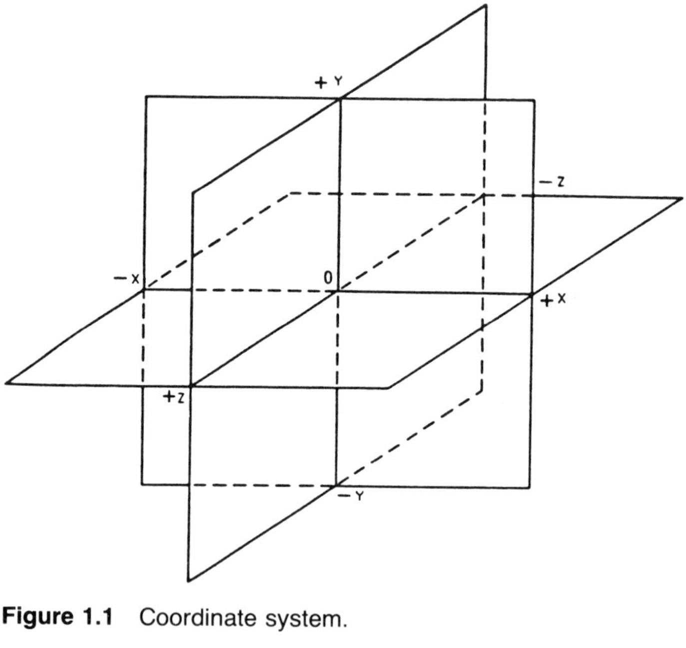

Figure 1.1 Coordinate system.

1.6 SPACE COORDINATION

In this section we shall see that the system of measurement described in section 1.5 can be used to define space in much the same way a mathematician describes points in a plane system of measurement. We shall see later in this text further similarities: (1) parametric descriptions of a line in space are quite similar to an equation of a line in a plane, (2) distance descriptions (true lengths) are a simple extension of the equation for the plane, (3) planes in space can be described by the location of three or more lines, and (4) solids can be described simply by means of equations.

For this section we will need to review the euclidean postulates for space coordination.

EP1 There is a set (group) of points (real numbers, mapped as ordered triplets). Certain subsets (members of the group) are called lines; at least two points are needed to define a line. Certain other subsets are called planes; at least three points are needed to define a plane, and solids contain at least four points not in any one plane.

EP2 There is exactly one plane containing any three noncolinear points.

EP3 Each plane satisfies all other postulates.

EP4 Each plane separates space into half-spaces with the following properties: (1) if two points are in the same half-space subset, a line between them does not intersect the plane, and (2) if two points are in different half-spaces, a line between them will intersect the plane.

From these few postulates all the theorems of modern descriptive geometry can be derived. However, we will not derive any theorems in this text. All that is required is an understanding of the elementary aspects of space geometry and the ability to visualize and display some space figures. In this regard, you should observe that the CAD display problem is one of representing three-dimensional relations in a plane, namely, the monitor face. Planes will be displayed by plotting a parallelogram, suggesting a portion of a plane. Lines that are behind planes as viewed in Figure 1.1 are dashed (hidden).

1.7 PROJECTIONS

Just as in section 1.6 we defined a one-to-one mapping of ordered pairs of real numbers onto the points of the plane, so for space projection we analogously define a one-to-one mapping of ordered triplets or real numbers onto reference planes. These projections depend on certain choices:

1. The projection of a point is the base of the perpendicular from the point to the reference plane as shown in Figure 1.2.
2. The projection of a line is the locus of the projections of all points of the line upon the reference plane.
3. The projection of a nonperpendicular line is a straight line as shown in Figure 1.3, while the projection of a perpendicular line is a point.

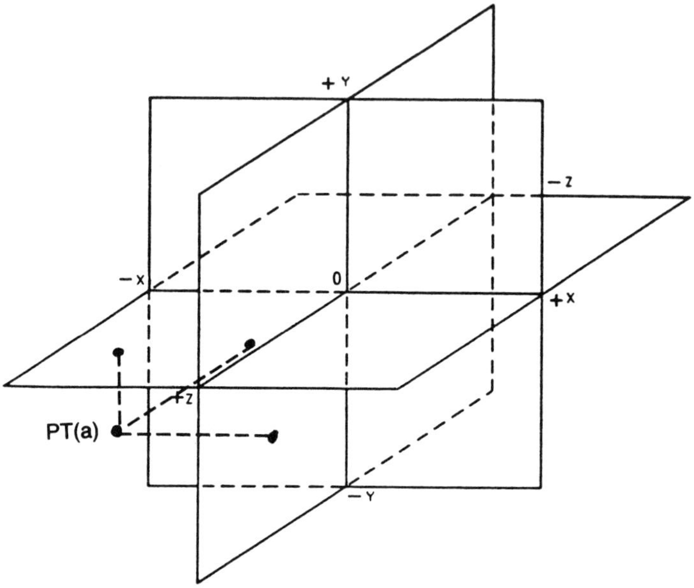

Figure 1.2 Projection of a point upon a reference plane.

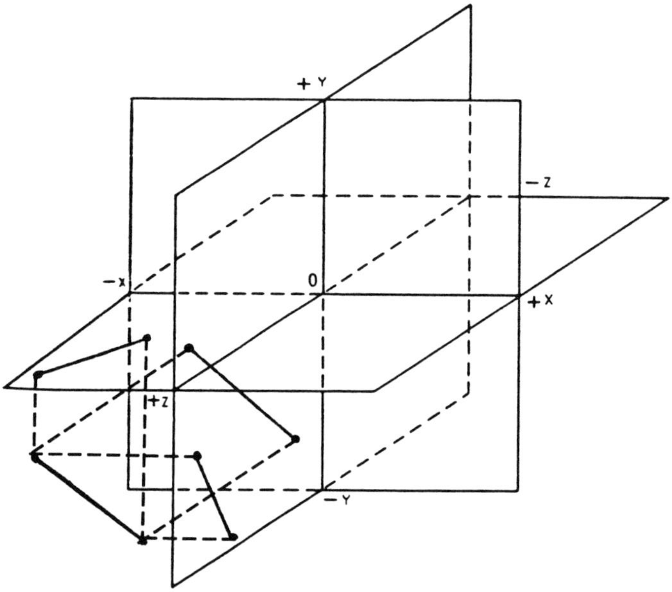

Figure 1.3 Projection of a line upon a reference plane.

4. The projection of a line parallel to a reference plane is a straight line on the reference plane and is equal in length to the given line (true length).
5. A projection of two lines such that a straight line joining any two of its points lies wholly between the two lines is called a projected plane.
6. If two points common to a plane are connected, the line lies wholly in that plane.
7. Points and lines lying in the same plane are said to be coplanar, while lines and points not sharing the same plane are called noncoplanar.
8. Lines common to the same plane must be either parallel or intersecting.

1.8 GEOMETRY CONVENTIONS

There are several conventions that are observed in coordinate space geometry. These conventions have to do with lines (parallel, intersecting, and perpendicular), planes, and solids. The following conventions apply:

Parallel Lines

Coplanar lines that cannot intersect however far projected are called parallel lines. Through a given point, only one line can be displayed parallel to another. Two coplanar lines are parallel if both are perpendicular to a third coplanar line. Two lines parallel to the same line are parallel to each other. Finally, two lines perpendicular to the same plane are parallel to each other.

Intersecting Lines

If two coplanar lines are not parallel, they must intersect. And two straight lines can intersect at only one point, called a concurrent point for both lines.

Perpendicular Lines

Lines that intersect at right angles are called perpendicular; an infinite number of space lines can be displayed perpendicular to a given line. However, from a given point only one line can be displayed perpendicular. Only one line can be displayed perpendicular to two intersecting lines.

Parallel Planes

Two planes are parallel if they cannot intersect. All of the conventions for lines apply to planes.

Intersecting Planes

If two planes intersect, one line is concurrent. This concurrent line is the locus of all points common to the two planes. If three planes intersect, the line of intersection is concurrent or parallel to each other. And if a plane intersects two parallel lines, the lines of intersection are parallel.

Perpendicular Planes

If a plane is perpendicular to a line in another plane, the two planes are perpendicular. If two planes are perpendicular, any line in one plane projected is perpendicular to any other.

Parallel Solids

The planes forming a solid are called its faces, and their concurrency is called the edge of the solid. If a face of one solid is parallel to another face of a second solid, the solids are said to be parallel.

Intersecting Solids

The intersection of two solids is a plane surface. This plane surface may be (1) a straight line and a point not in the line, (2) three points not in a straight line, or (3) two parallel lines.

Perpendicular Solids

If a plane in one solid is perpendicular to a plane in another solid, the solids are considered to be perpendicular. In the case of intersecting solids, the dihedral angle between the faces of the solids must be 90 degrees.

1.9 POINTS IN SPACE

We have defined euclidean space and have given it coordinates. But often in the applications of mathematics to problems of engineering and computer science it is the converse situation that prevails. We are given ordered triplets of real numbers and use space to illustrate graphically the relations between the ordered triplets. In some cases there is no need to choose the display option; often the CAD workstation has only one method for graphic display. In most systems a point (PT) is simply

$$\mathrm{PT}(a_1, a_2, a_3)$$

Furthermore, there is no definite way to orient the display axes, but it is customary to default the CAD display so that X is horizontal, Y is vertical, and Z is in and out (perpendicular) of the display screen. For observation purposes you may position and reposition the display axes as demonstrated in Figure 1.4. The three coordinate planes shown in Figure 1.1 separate display space into eight octants. Usually only one of them is given a name and referenced in mathematics; this is the first octant, which consists of the sets of all points having positive coordinates only, and this is stated as

$$\mathrm{PT}(x_1, y_1, z_1)$$

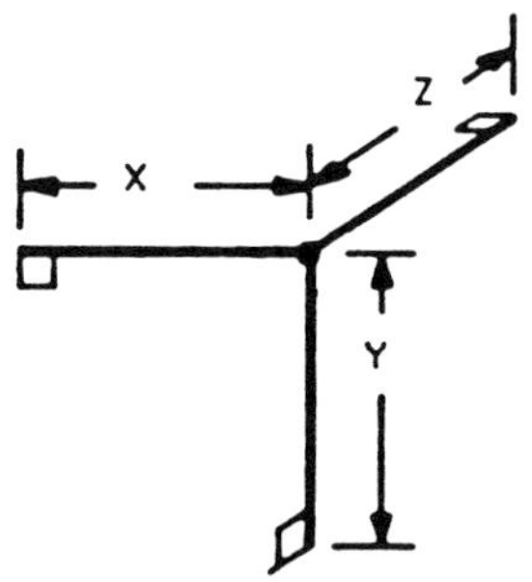

Figure 1.4 Point definition in *X*, *Y*, and *Z*.

1.10 DISTANCE BETWEEN POINTS

We are provided the two points PT(1) = (X(1),Y(1),Z(1)) and PT(2) = (X(2),Y(2),Z(2)). We choose to add point 3 in Figure 1.5 so that points 1 and 3 have the same Z value. The distance | PT1PT3 | is given by the equation in the plane, namely

$$| \text{PT1PT3} | = ((X(1) - X(2))^{**}2 + (Y(1) - Y(2))^{**}2)^{**}.5 \quad \textbf{(1)}$$

The points PT2 and PT3 are on a straight line and

$$| \text{PT2PT3} | = ((Z(1) - Z(2))^{**}2)^{**}.5 \quad \textbf{(2)}$$

This makes PT1PT2PT3 a right triangle, and by pythagorean theorem,

$$| \text{PT1PT2} | = ((X(1) - X(2))^{**}2 + (Y(1) - Y(2))^{**}2 + (Z(1) - Z(2))^{**}2)^{**}.5 \quad \textbf{(3)}$$

1.11 DIRECTION OF LINES

In the remaining chapters we will be using a computer-aided graphics package designed for descriptive geometry users. For this purpose we need to define direction angles of a line in space. We begin by letting any line (L) have starting coordinates of X(0),Y(0),Z(0) and ending coordinates of X(1), Y(1), Z(1). Next determine the point distance (D). The direction angles are set as (1) the X-direction cosine, (2) the Y-direction cosine, and (3) the Z-direction cosine. If we label these angles $C1$, $C2$, and $C3$ respectively, then the following equations can be used by the display software

$$C1 = \frac{X(1) - X(0)}{\text{D}}$$

$$C2 = \frac{Y(1) - Y(0)}{\text{D}}$$

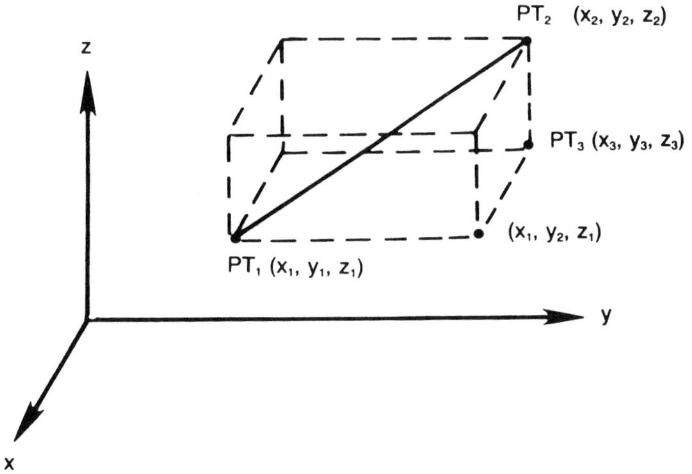

Figure 1.5 The distance formula.

$$C3 = \frac{Z(1) - Z(0)}{D}$$

If $C1$, $C2$, $C3$ are the direction cosines of a line, then
$C1^{**}2 + C2^{**}2 + C3^{**}2 = 1$

1.12 PARAMETRIC EQUATIONS OF LINES

The remaining chapters also use a computer-aided graphics package designed with parametric equations of lines. There are two systems associated with the parametric equation of a line in space. One is the directed distance just presented where

$$C1 = \frac{X(1) - X(0)}{D} \qquad \textbf{(1)}$$

$$C2 = \frac{Y(1) - Y(0)}{D} \qquad \textbf{(2)}$$

$$C3 = \frac{Z(1) - Z(0)}{D} \qquad \textbf{(3)}$$

The other is the rectangular coordinate system represented earlier by the ordered triplet (X,Y,Z).

Given a directed line with cosines $C1$, $C2$, and $C3$, let $X1$, $Y1$, and $Z1$ be a point on the line at distance D then
$X = X1 + C1^*D$
$Y = Y1 + C2^*D$ and
$Z = Z1 + C3^*D$

1.13 POINT AND LINE DEFINITION

Chapter 2 of this book explains how computer-aided constructive geometry enables the user to (1) display, store, reproduce, modify, or transmit CADD elements of any object involving form so that the output will reveal all the essential facts; (2) determine from (1) any and all desired facts as to distances, angles, intersections, areas, volumes, and mass properties; and (3) generate from (1) and (2) any constructions required to introduce given distances, angles, intersections, areas, volumes or mass properties as design data. The purposes of computer-aided constructive geometry are accomplished through the employment of five CADD principles: (1) orthographic projection, (2) parallelism, (3) perpendicularity, (4) intersections, and (5) revolution.

1.14 PLANE DEFINITION

Chapter 3 of this book details the various types of plane surfaces used in descriptive geometry. A plane thus defined can be displayed by displaying any two lines within the plane surface. The CADD display screen completely describes the projections of these lines on the coordinate planes. All lines lying in a given plane surface are called *coplanar* lines. The significant qualities of such lines are that they either intersect or are parallel. Since every plane in computer-aided constructive geometry can be extended to infinity, in theory, every displayed plane surface therefore may be extended until it intersects at least two of the three coordinate planes; these intersections are called *principal traces*. If two plane surfaces intersect each other, they also leave a trace along the line of intersection.

1.15 SOLID DEFINITION

Chapter 4 of this book describes an interactive CADD program based on the Rylander software approach. This approach allows the construction of 3-D images from combinations of elementary shapes such as *sphere, cylinder,* and *toroid*. For these purposes, the author has added cone and pyramid as basic primitives and used Rylander's *polymesh* technique to render *joined* or assembled objects—solids.

The options available for solid presentation include

1. Wireform in any axonometric orientation, random shaded polygon surfaces, or halftone presentations,
2. Solid models with hidden line removal in any axonometric rotation, with color shading in random or halftone patterns, and
3. Light source for realistic model shades and shadows to include basic primitives for *cube, cone, pyramid, sphere,* and *torus*. With these building blocks, realistic models have surprising detail as shown in Figure 1.6.

1.16 POINT/LINE/PLANE RELATIONSHIPS

Chapter 5 of this book contains the working relationships for constructive geometry, namely, the point, line, and plane combinations that can be displayed in engineering problem solving. Both principal and plane tracing are used throughout the many CADD displays shown as figures in this chapter.

1.17 POINT/LINE/PLANE ANALYSIS

Chapter 6 of this book covers the concepts of revolution for (1) true length of a line, (2) distance from a point to a line, (3) distance between two parallel planes, (4) true distance, and (5) common perpendiculars. Common engineering problems are displayed through the CADD format.

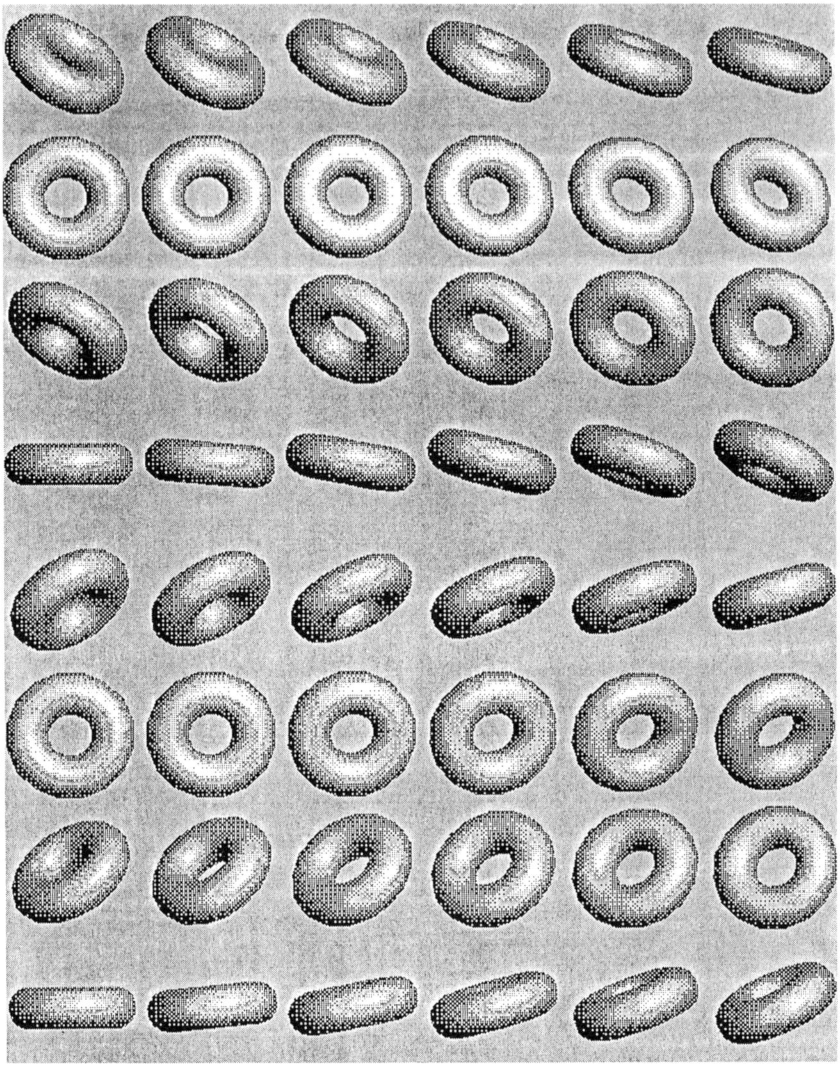

Figure 1.6 Solid modeling output of a torus.

1.18 MEASUREMENT OF PLANE SURFACES

Chapter 7 of this book deals with true shape of plane surfaces, true angle between surfaces, the bisector of angles between two intersecting planes, and several techniques for plane surface analysis.

1.19 PLANES AND SOLIDS

Chapter 8 of this book introduces the classification of planes and solids. Several general types of planes are covered: (1) straight, (2) curved, (3) doubled curved,

(4) conics, (5) spirals, and (6) helices. Several types of solids are covered: (1) sphere, (2) ellipsoid, (3) torus, (4) cylinder, and (5) cone.

1.20 DEVELOPMENT AND INTERSECTIONS

Chapters 9, 10, and 11 deal with tangent planes and solids, intersection of planes and solids, and development of solids using surfaces, respectively. Each chapter presents CADD screen displays showing construction, solution, and analysis of common descriptive geometry problems.

1.21 CHAPTER PROBLEMS

The problems appearing at the end of the remaining chapters have been selected to introduce the basic types of broad, general problems encountered by an engineer or engineering technician using a CADD display system. These problems emphasize the interdisciplinary aspects of constructive geometry, which require a number of related pieces of CADD software. The software used for the production of figures in this text can be ordered from Autodesk, Inc.

2 Point and Line Definition

Computer-aided constructive geometry is that division of engineering science that enables the user to (1) display, store, reproduce, modify, or transmit CADD elements of any object involving form so that the output will reveal all the essential facts; (2) determine from (1) any and all desired facts as to distances, angles, intersections, areas, volumes, and mass properties; and (3) generate from (1) and (2) any constructions required to introduce given distances, angles, intersections, areas, volumes, or mass properties as design data. The purposes of computer-aided constructive geometry are accomplished through the employment of five CADD principles: (1) orthographic projection, (2) parallelism, (3) perpendicularity, (4) intersections, and (5) revolution.

2.1 POINT AND LINE RELATIONSHIPS

The engineering problems of constructive geometry are related to 3-D objects. They can be solved on any CADD system through the medium of plane or 2-D geometry presented at the CRT face or through printed matter on paper. Such a set of CADD procedures must therefore be employed to reduce the 3-D problems to a 2-D display format. It is obvious that the process of description must produce a group of related images in which the following data types are presented: (1) front view = X,Y; (2) top view = X,Z; and (3) profile view = Z,Y. By using any two views, all three dimensions can be studied.

The required constructive geometry descriptions are called projections and are made upon a set of conventional intersecting planes as shown in Chapter 1 (Fig. 1.1). The projections are made upon these planes by means of points (Fig. 1.2), or lines (Fig. 1.3) through all significant data elements of any structure and perpendicular to the respective planes of description. This concept is called orthographic projection.

2.2 PLANES OF PROJECTION

The three conventional planes are called horizontal (top), frontal (front), and profile (side). These planes are also used as coordinate planes for the display points. For convenience, we can separate these planes by the use of a reference line. The reference line between the horizontal and frontal planes of projection is labeled H/F. The line of intersection between the frontal and profile planes is labeled F/P.

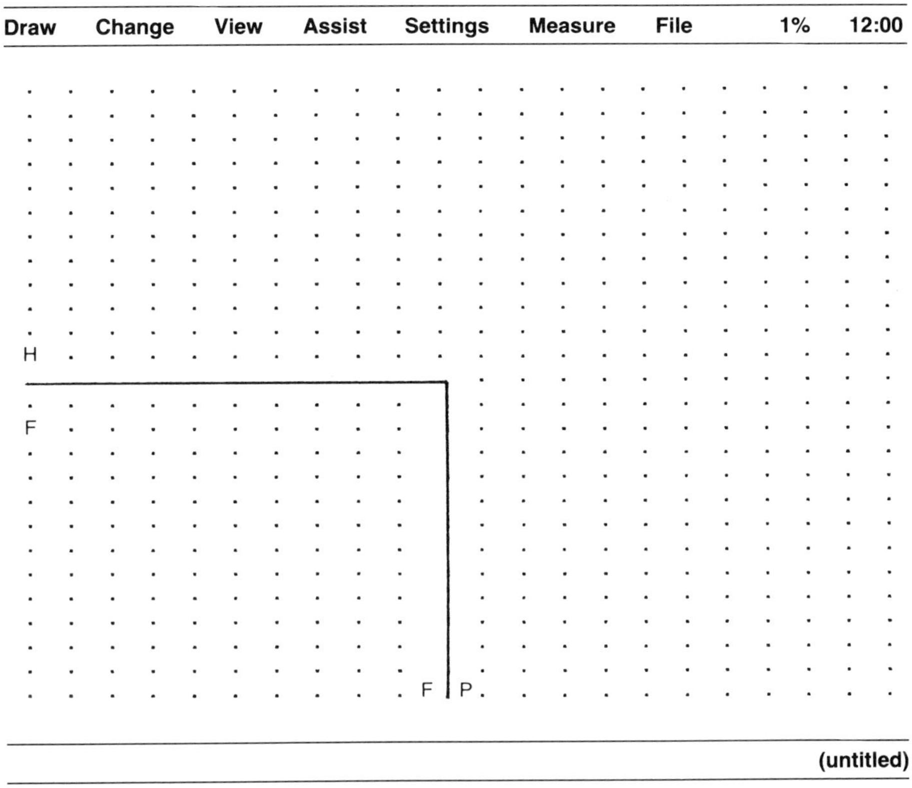

Figure 2.1 *H/F* and *F/P* reference lines.

2.3 PROJECTIONS OF A POINT

Figure 2.1 represents a CADD display of the reference lines *H/F* and *F/P*. Compare, for example, Figures 1.2 and 2.2. In Figure 1.2 a point is shown as its data type in relation to the coordinate locations. In Figure 2.2, the data is mapped onto the projection planes as constructive point *A(h)*, *A(f)*, and *A(p)*. The representations or descriptions of points and lines on the planes of projection are called views.

The *H* view of point *A* in Figure 2.2, for example, is the point in which the orthogonal line of sight through *A* pierces the *H/F* reference line. This line of sight is infinite in extent but there is no good reason for describing any but the portion that lies on the grid background. A projection line may be located on a grid or off the grid background, and it passes through the data point perpendicular to the reference lines.

2.4 LINE OF SIGHT FOR A POINT

The projection of any data point, say point *A*, has *X*, *Y*, and *Z* values. Referring to Figure 2.2. again, notice that point *A* is given in the frontal view by its *X* and *Y* values. If *Z* is known, it can be located in the top and side views by projection.

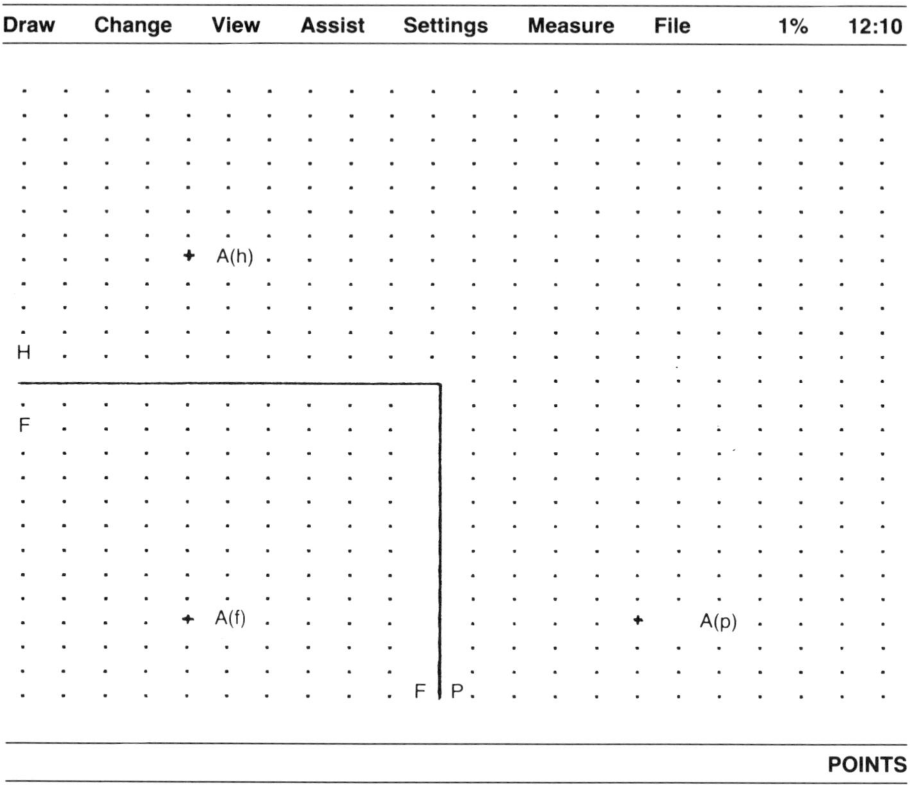

Figure 2.2 Constructive point *A*.

A line of sight is displayed perpendicular to *H*/*F* as shown in Figure 2.3. Notice that the series of arrows are for explanation only and do not normally appear on a working problem.

2.5 LOCATION OF POINTS

Points are located by *X*, *Y*, and *Z* coordinates and displayed for viewing. In some cases one of the coordinates *X*, *Y*, or *Z* may not appear and can be located as shown in Figure 2.4 where case 1 is located by projecting into the top and side views with *Z* coordinate. Case 2 is located by projecting from side and top into the frontal view, and case 3 is projected in the frontal with coordinate *X* and then into the top view.

2.6 LINES

Since a line is generated in CAD displays by a point moving in some fixed direction, the *H* projection of a straight line is made up of the *H* projections of the successive positions (*X* and *Z*) of the generating point. Likewise the *F* projection is made up of the *X* and *Y* positions, and the *P* projections are made up of the *Z* and *Y* positions.

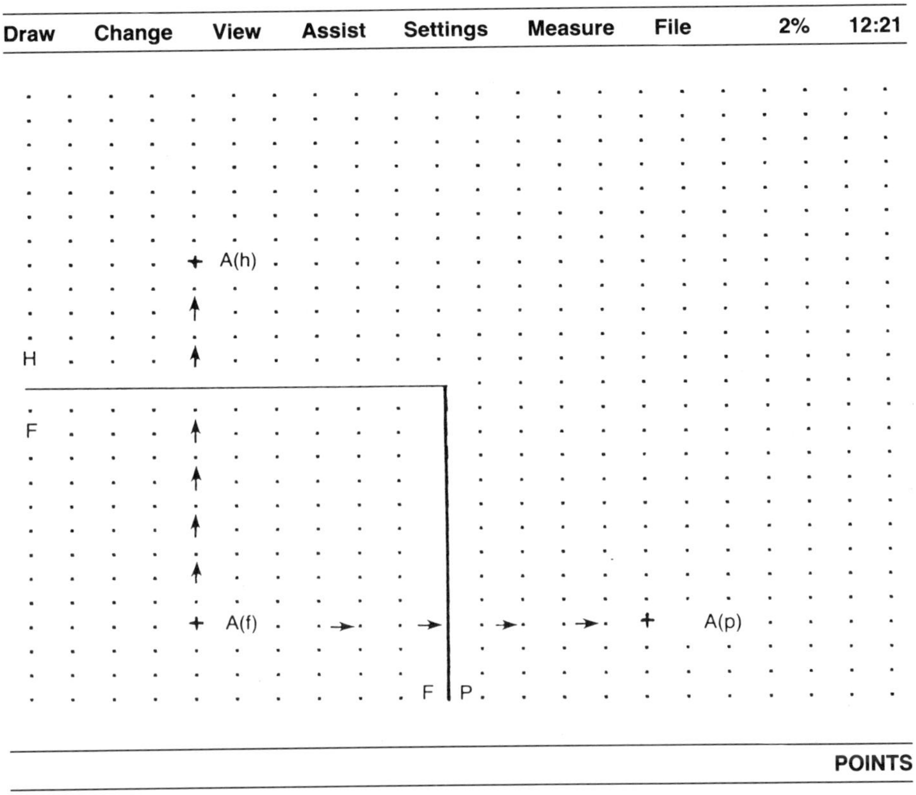

Figure 2.3 Line of sight for point *A*.

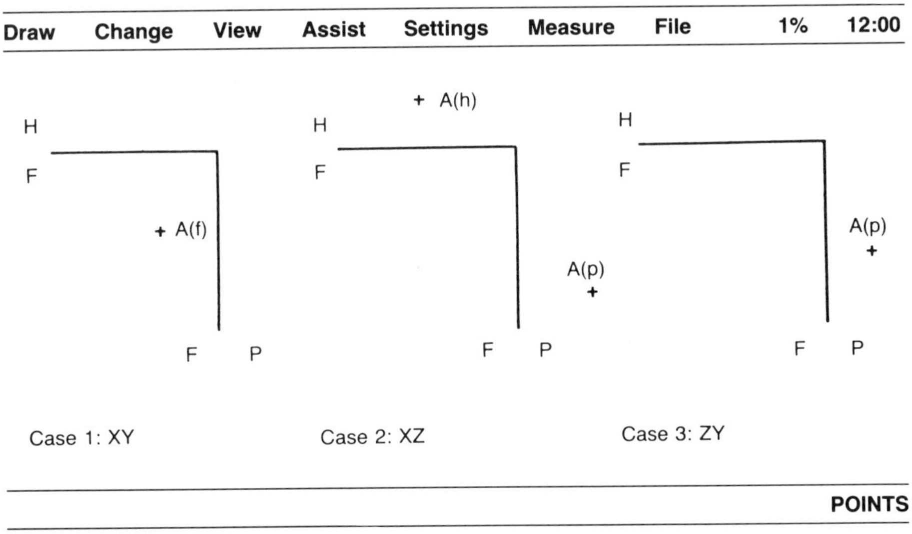

Figure 2.4 Point locations.

In this text it may be assumed that the term *line* implies a straight line unless otherwise noted. Since any two generating points determine a line, a line is usually designated by its two extremities if it has a fixed length. Otherwise any two points of the line may be chosen at random for the purpose of locating the entire line in any other view. In general, the line is read from left to right. Lines in a profile plane of projection are read from front to rear. Notice in Figure 2.5 that two sets of line segments *(AB)* and *(CD)* are displayed. The direction of line *AB* is left to right, while line segment *CD* intersects the *H/F* reference line. Line *AB* is called a plane line and line *CD* is called a line with a trace point. The point at which line *CD* pierces the *H* plane of projection is called a trace point. Trace points are often chosen as ending points for line segments, since the continuation of the line is useless for most CAD operations.

2.7 OBLIQUE LINES

Lines can also be classified like point locations, in that they may be parallel to a reference line *(H/F)*, perpendicular, or neither of these. Lines that are neither parallel nor perpendicular are called oblique line segments. Line segment *AB* in Figure 2.5 is called a principal line, while line *CD* is an oblique line.

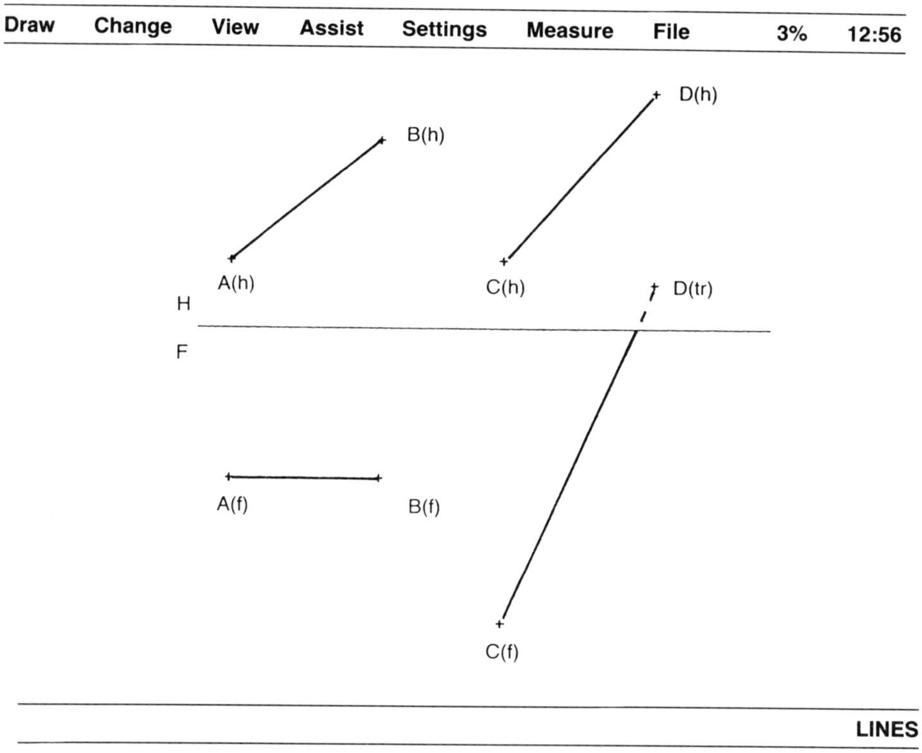

Figure 2.5 Line locations.

2.8 PRINCIPAL LINES

Line segment *AB* in Figure 2.5 is parallel to *H/F* reference line and appears in true length in the horizontal view. Therefore, this line is called a *horizontal* line in addition to a principal line and a true length line. Principal lines may appear true length in any view and are called horizontal, frontal, or profile depending upon the orientation of the parallel relationship. If a line is parallel or perpendicular to a reference line, it will appear in true length in the adjacent view for parallel and as a point view for perpendicular.

2.9 LOCATION OF LINES

If two projections of a line are given, the line is in general completely determined. Figure 2.6 shows the three classic cases for line images.

Case 1, a frontal/horizontal line, is solved in Figure 2.7 by projecting into the profile to establish a point view *B,A(p)*. Case 2, an oblique line, is solved by projection into the front view, and case 3, a frontal line, is solved by projection into the top view. Each true length is labeled *TL* in each case and the direction of sight is displayed with ↑ for *H/F* and < for *F/P*.

2.10 LOCATION OF POINTS AND LINES

Lines are composed of an infinite number of points. The solution of constructive geometry problems requires that the locations of specific points on lines in space be determined. Figure 2.8 shows three cases for point locations.

In case 1, any point shown on either line view may be projected to locate the same point. Projection is always available in this manner. In case 2, the intersection of the lines is tested by point projection. Notice that the intersection of one line view is the same in the next view. Therefore, the lines intersect. In case 3,

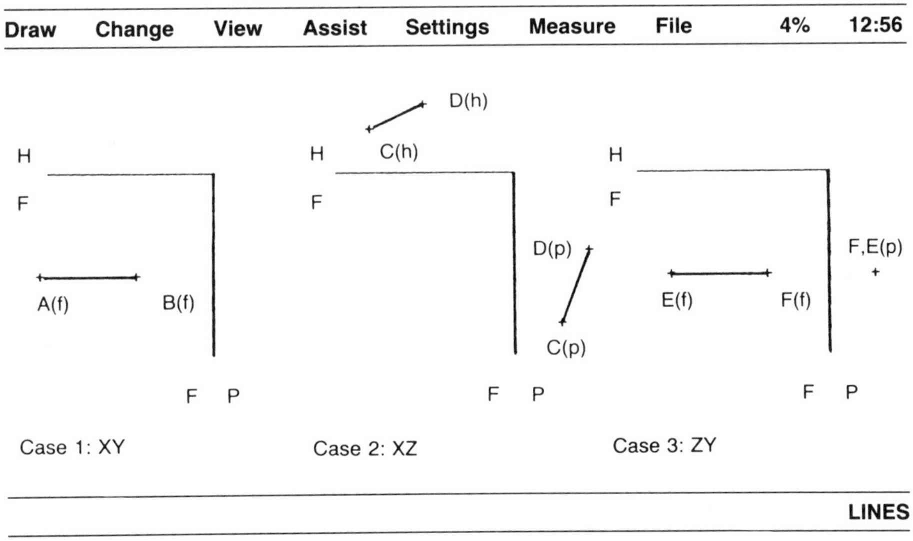

Figure 2.6 Line locations.

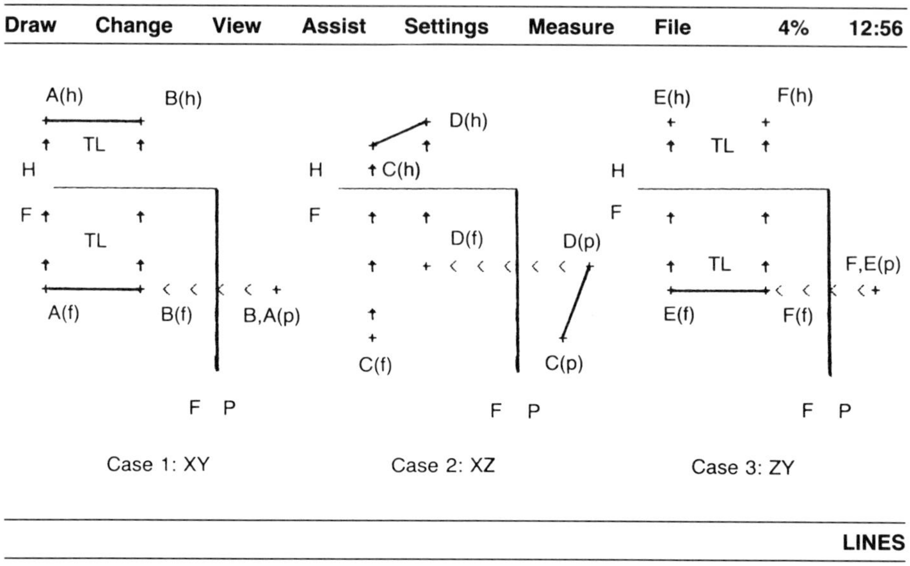

Figure 2.7 Line solutions.

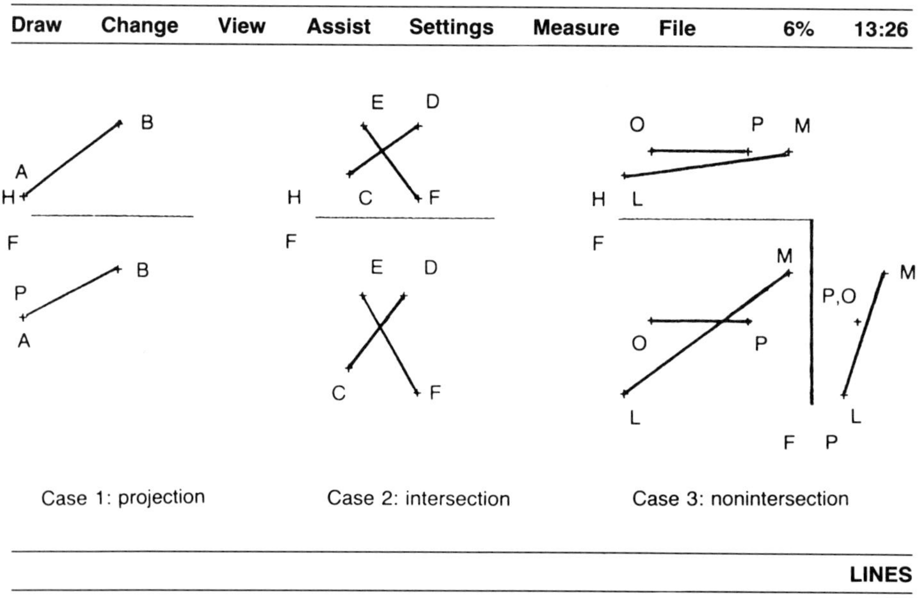

Figure 2.8 Line and point locations.

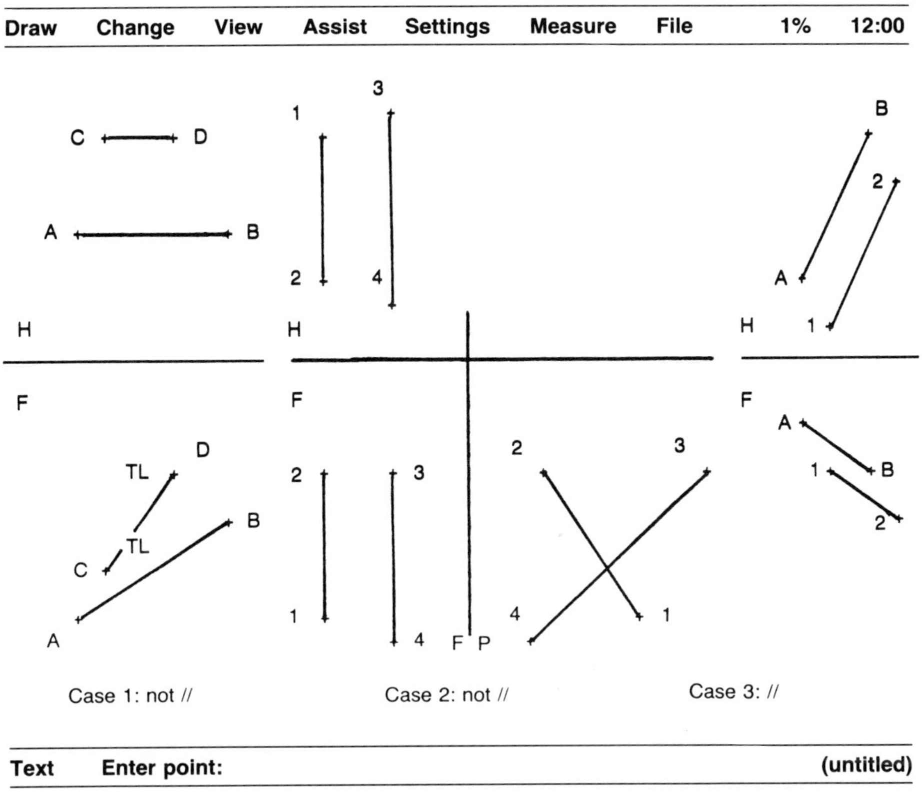

Figure 2.9 Parallel lines.

there is no intersection. Point projection is not needed in this case because lines do not appear to intersect in the top or profile views. Oblique line projections should always be tested for intersection with the point projection method. If the intersections are tested to be directly above and below each other, then the lines intersect.

2.11 PARALLEL LINES

The testing method used for points and lines is valuable for determining whether or not oblique lines are parallel to each other. While it is true that *principal* lines that appear to be parallel in two views are parallel in all others, but is this true for oblique lines? Figure 2.9 shows three particular cases. In case 1, line *AB* appears to be parallel to *CD* in the top view but is not parallel in the frontal plane and no testing is needed. In case 2, lines 1,2 and 3,4 appear to be parallel in both the top and front but are clearly not when projected into the profile. Here the test was the construction of the profile view. In case 3, the lines appear parallel in top and front and are parallel after testing or projection into a profile view. Display lines are parallel to each other when all tests prove parallel relationships.

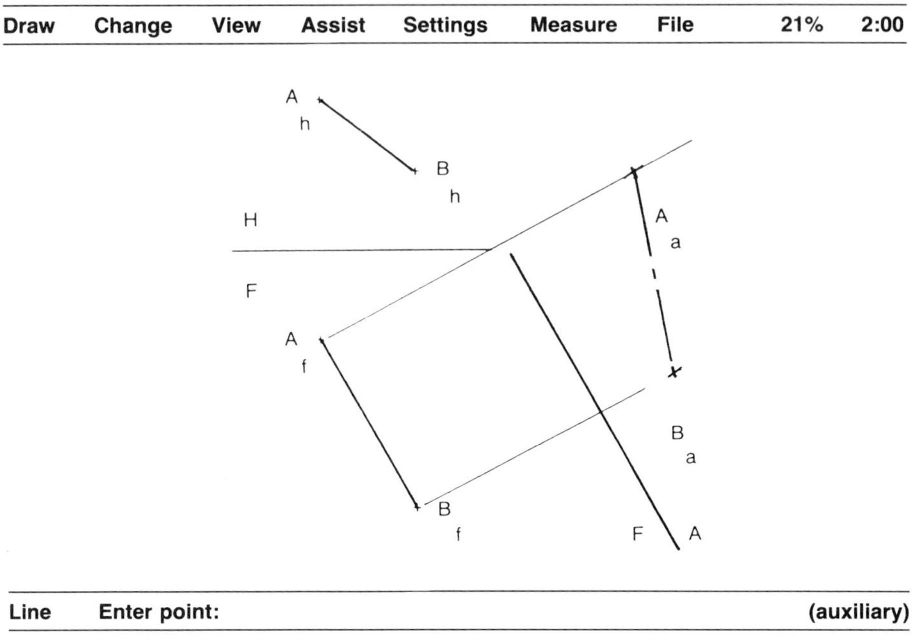

Figure 2.10 True length by auxiliary projection.

2.12 PARALLEL PROJECTION

Lines may be parallel to other display lines or they may be parallel to a reference line. In order for a line to be a principal line, one view must appear parallel to *H/F* or *F/P*. Because principal lines can be measured true length (TL), they are preferred in most constructive geometry problems. Many times an oblique line can be displayed in true length by the use of a parallel reference line. Figure 2.10 uses a parallel reference line labeled *F/A* to display the line image in its true length. This method is called auxiliary projection.

2.13 PERPENDICULAR LINES

If two lines intersect and make a 90-degree angle with each other, they are said to be perpendicular. If a line is displayed parallel to a reference line *(H/F),* it is also perpendicular to a reference line *(F/P).* If two perpendicular lines are both parallel to the *H/F* as shown in case 1, Figure 2.11, the frontal view will also appear perpendicular. In case 2 we see that any number of cases can appear perpendicular in the front view, while case 3 is a 3-D model of the perpendicular line method.

2.14 GEOMETRIC CHARACTERISTICS OF A LINE

The geometric characteristics of a line are determined by placing the line inside a coordinate model as shown in case 1 of Figure 2.12. Here the *X* (horizontal), *Y* (vertical), and *Z* (profile) coordinates for the end points of the line are displayed. The intersection of the *X,Y* planes has been labeled *BL* in case 2 and is called the base line of the geometric model. Line *AB* is perpendicular to and intersects the

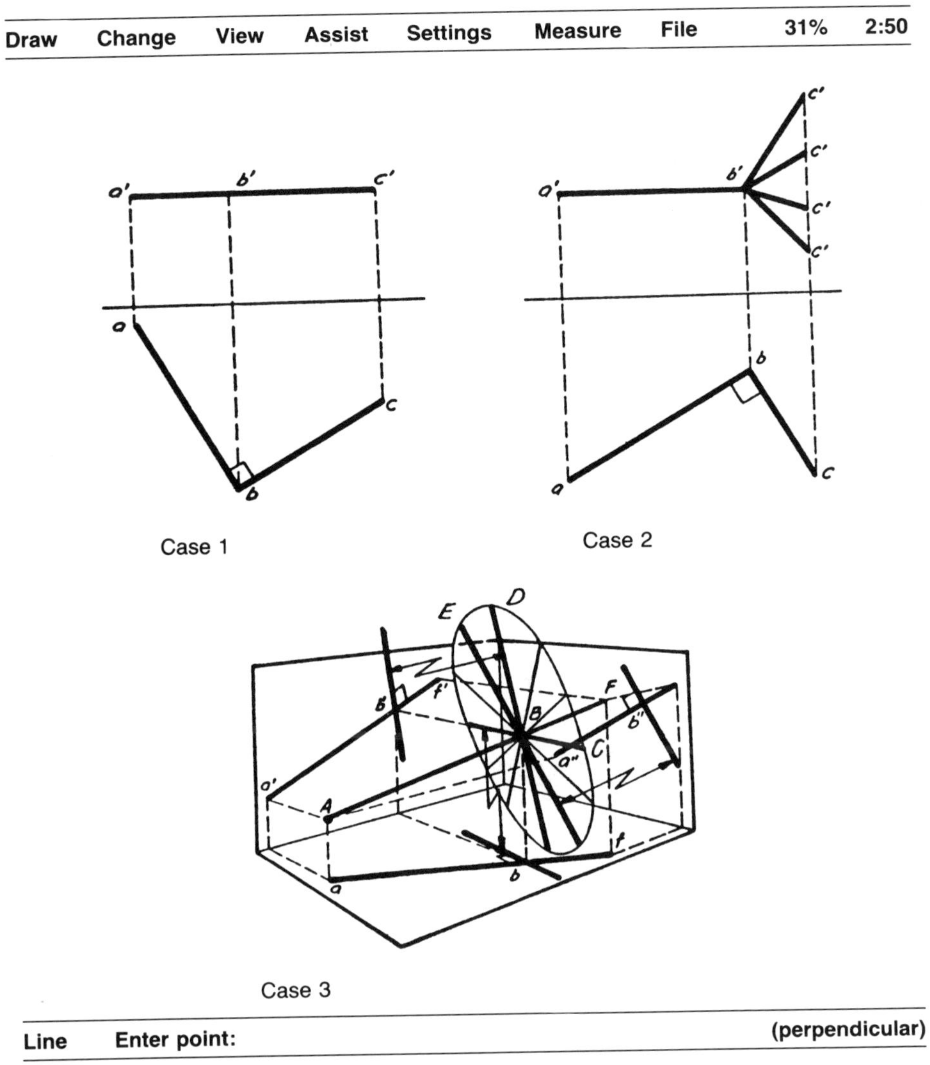

Figure 2.11 Perpendicular line locations.

horizontal plane because of its end point coordinates. For this reason a direct projection across an *H*/*F* reference line is impossible.

2.15 LINES PERPENDICULAR TO *F* OR *V*

The geometric characteristics of a perpendicular line are determined by placing the line inside a coordinate model as shown in case 1 of Figure 2.13. Here the coordinates for the end points of the line are displayed. *BL* in case 2 is displayed parallel to the line segment *CD*, which makes the line perpendicular to the vertical planes *F* and *P*. Line *CD* is perpendicular to and intersects *PL* (profile line). Both lines *AB* and *CD* are true length lines called normal or principal lines.

Draw Change View Assist Settings Measure File 12% 3:20

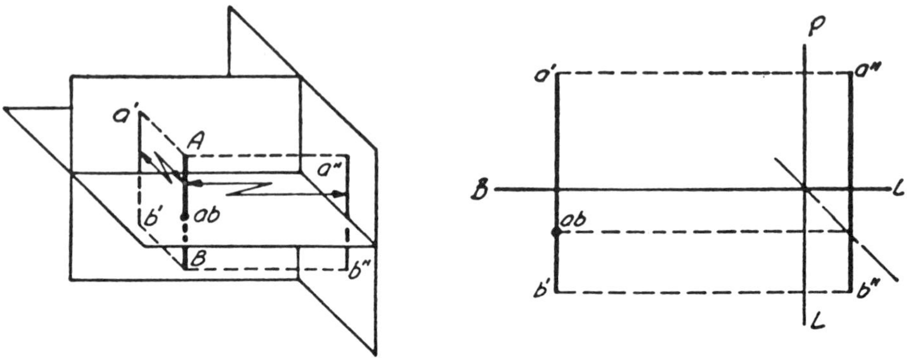

Case 1 Case 2

Line Enter point: (modeling)

Figure 2.12 Line perpendicular to *H*.

Draw Change View Assist Settings Measure File 14% 2:30

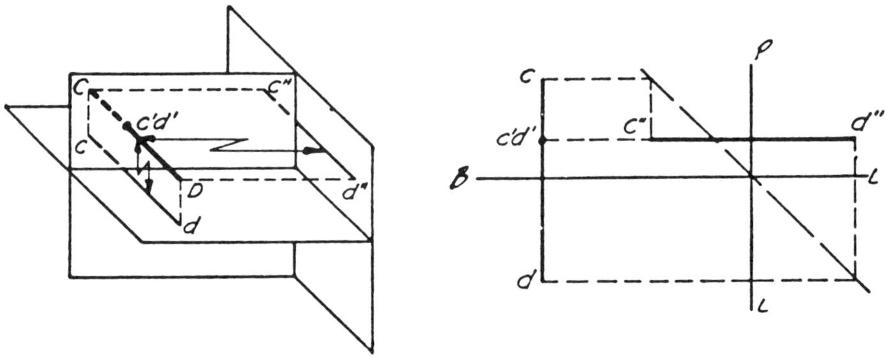

Case 1 Case 2

Line Enter point: (perpline)

Figure 2.13 Line perpendicular to *V*.

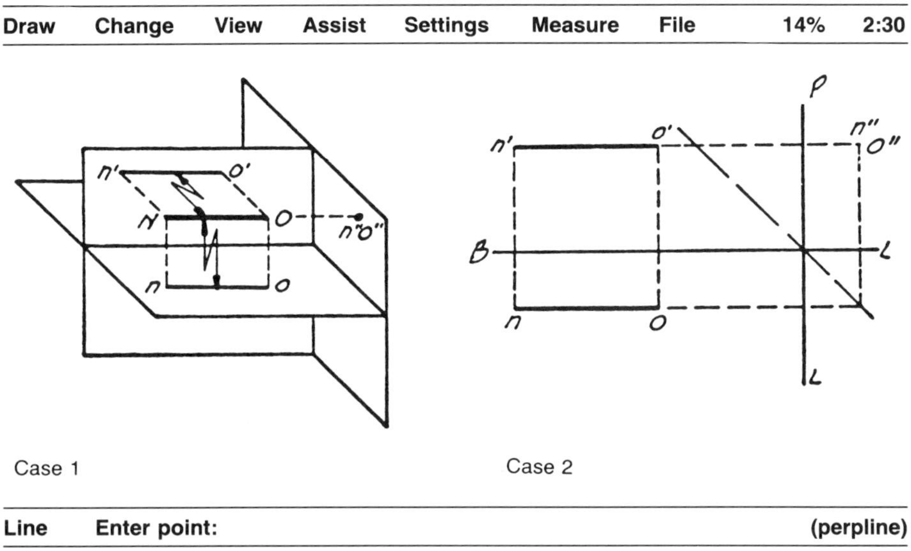

Figure 2.14 Line perpendicular to *P*.

2.16 LINES PERPENDICULAR TO *P*

The characteristics of a perpendicular line are displayed by placing the line inside the model shown in case 1 of Figure 2.14. Here the coordinates for the end points of the line *NO* are displayed. *BL* and *PL* in case 2 are displayed parallel and perpendicular to the line segment *NO*, which makes the line perpendicular to the profile plane *P*. Line *NO* is perpendicular to *PL* (profile line). Lines *AB*, *CD*, and *NO* are true length lines called normal or principal lines.

2.17 LINES PARALLEL TO *P* INCLINED TO *H* OR *V*

The characteristics of parallel lines can be displayed by placing them inside the model shown in case 1 of Figure 2.15. Here the coordinates for the end points of the line *EF* are displayed. *BL* and *PL* in case 2 are displayed in relationship to the line *EF*. The angle that line *EF* makes with *H* is called its slope angle. The trace *e″f″* is in true length and the line is called a profile line. Lines *AB*, *CD*, *NO*, and *EF* are all true length lines.

2.18 LINES PARALLEL TO *H* INCLINED TO *P* OR *V*

The characteristics of parallel lines can be displayed by placing them inside the model shown in case 1 of Figure 2.16. Here the coordinates for the end points of the line *GK* are displayed. *BL* and *PL* in case 2 are displayed in relationship to the line *GK*. The angle that line *GK* makes with *V* is called its bearing angle. The trace *g′f′* is in true length and the line is called a horizontal line. Lines *AB*, *CD*, *NO*, *EF*, and *GK* are all true length lines.

Draw **Change** **View** **Assist** **Settings** **Measure** **File** **11%** **1:20**

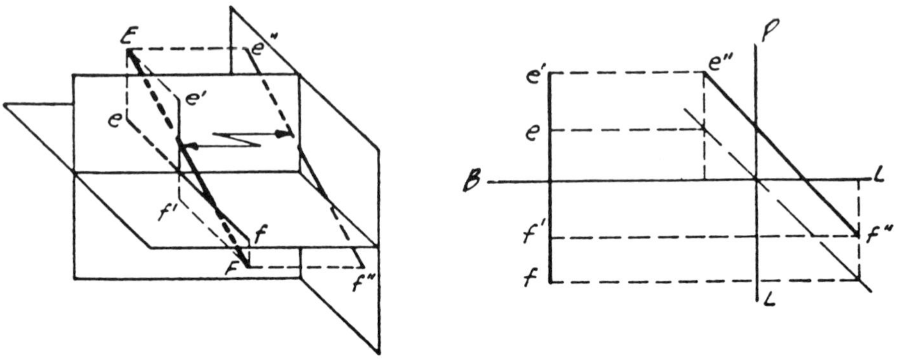

Case 1 Case 2

Line **Enter point:** **(paraline)**

Figure 2.15 Line parallel to *P*.

Draw **Change** **View** **Assist** **Settings** **Measure** **File** **15%** **2:40**

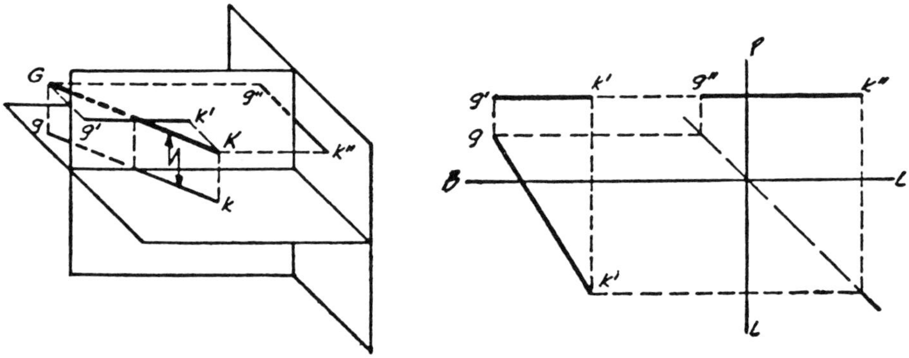

Case 1 Case 2

Line **Enter point:** **(paraline)**

Figure 2.16 Line parallel to *H*.

Draw Change View Assist Settings Measure File 13% 3:40

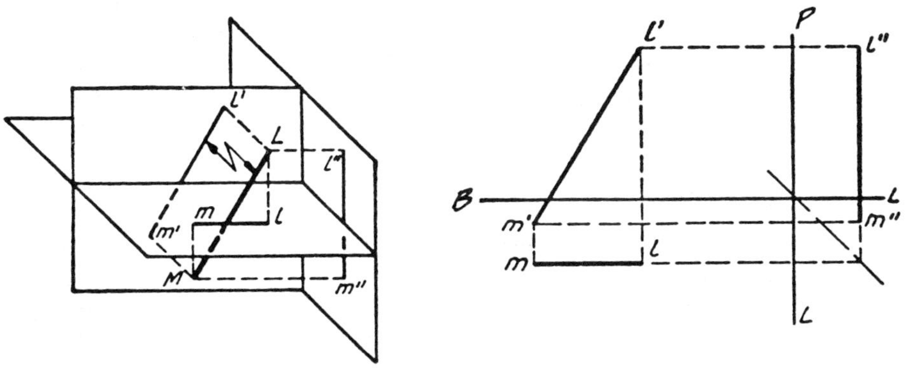

Case 1 Case 2

Line Enter point: (paraline)

Figure 2.17 Line parallel to *V*.

Draw Change View Assist Settings Measure File 17% 4:22

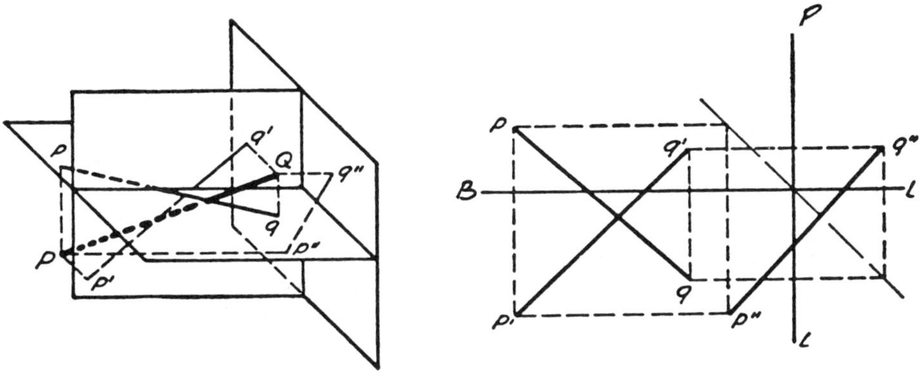

Case 1 Case 2

Line Enter point: (obliques)

Figure 2.18 Line inclined to *HVP*.

2.19 LINES PARALLEL TO *V* INCLINED TO *H* OR *P*

The characteristics of parallel lines can be displayed by placing them inside the model shown in case 1 of Figure 2.17. Here the coordinates for the end points of the line *LM* are displayed. *BL* and *PL* in case 2 are displayed in relationship to the line *LM*. The angle that line *LM* makes with *P* is called its dip angle. The trace $l'm'$ is in true length and the line is called a vertical or frontal line. Lines *AB, CD, NO, EF, GK,* and *LM* are all true length lines.

2.20 LINES INCLINED TO *HVP*

The characteristics of nonparallel lines can be displayed by placing them inside the model shown in case 1 of Figure 2.18. Here the coordinates for the end points of the line *PQ* are displayed. *BL* and *PL* in case 2 are displayed in relationship to the line *PQ*. The angles that line *PQ* makes with *H, V,* or *P* are called oblique angles. The traces pq, $p'q'$, and $p''q''$ are not in true length and are called oblique traces. Lines *AB, CD, NO, EF, GK,* and *LM* were all true length, however.

2.21 CHAPTER PROBLEMS

The problems appearing at the end of this chapter have been selected to introduce the basic type of point and line problems encountered by an engineer or engineering technician using a CADD display system. These problems emphasize the interdisciplinary aspects of geometric construction based upon a coordinate model introduced in Chapter 1 of this text.

PROBLEM SET 1 *Points*

In each problem display the point defined.

1.1 *A(f)* (Fig. 2.2)
1.2 *A(h)* (Fig. 2.2)
1.3 *A(p)* (Fig. 2.2)
1.4 *LS* (Fig. 2.3)
1.5 *LS* (Fig. 2.3)
1.6 *A(h)* (Fig. 2.4[1])
1.7 *A(p)* (Fig. 2.4[1])
1.8 *A(f)* (Fig. 2.4[2])
1.9 *A(h)* (Fig. 2.4[3])
1.10 *A(f)* (Fig. 2.4[3])

PROBLEM SET 2 *Line definition*

In each problem display the line defined.

2.1 Coord. (Fig. 2.5)
2.2 *P* plane (Fig. 2.6[1])
2.3 *F* plane (Fig. 2.6[2])
2.4 *H* plane (Fig. 2.6[3])
2.5 Parallel (Fig. 2.9)
2.6 Principal (Fig. 2.9)
2.7 Auxiliary (Fig. 2.10)
2.8 Perpendicular (Fig. 2.11)

3 Plane Definition

A plane is defined by any two lines in it and is completely described by the projections of these lines on the coordinate model shown in Chapter 1, Figures 1.1 through 1.3. All lines lying in a given plane are called *coplanar* lines. The significant qualities of such lines are that they either intersect or are parallel as shown in Chapter 2. Since every plane used in this chapter must intersect at least two of the coordinate plane surfaces shown in Figure 1.1, these intersections may serve as the "any two lines," and their use in this chapter serves to simplify the following discussions of the relationships of points, lines, and planes. For these reasons, we will refer to the lines of intersection of planes and the coordinate model surfaces as *traces*. To keep the coordinate plane surfaces of the model separate in discussions, we will call the coordinate planes *H (X,Z)*, *V (X,Y)*, and *P (Z,Y)* as displayed in Figure 3.1.

The display pattern for a plane trace is a medium dash, dot, long dash, dot, medium dash as shown in Figure 3.1. Each display plane must have a name, a single letter (i.e., *R*), or the notation of the points representing the plane (i.e., *ABC*) as used in Chapter 2. In Chapter 2, section 2.2, planes of projection (*H*, *F*, and *P*) were used to directly project object points; a plane surface may be displayed using this technique. If, however, a display plane is described by its traces, the traces may be designated as *HR*, *VR*, and *PR*.

3.1 PLANES OF PROJECTION OR TRACES ON COORDINATE PLANES

The intersection of the *H* and *V* planes of the coordinate model is labeled *BL* in Figure 3.2 and is known as the base line. The intersection of the *P* coordinate plane with the *V* coordinate plane is known as the profile line and is labeled *PL* in Figure 3.3. The three conventional planes introduced in the preceding chapter are named the *planes of projection*, and they are called *H*, *F*, *P* (horizontal, frontal, and profile). These reference planes are locked into a traditional glass box foldout position based upon the third quadrant as shown in Figure 1.3. The three conventional planes introduced in Figure 1.1 are called *coordinate planes*, and they are called *H*, *V*, *P* (horizontal and vertical). These three mutually perpendicular planes serve as the coordinate planes for the description of points, lines, and planes displayed by the Autodesk software used with this textbook. The coordinate planes are *not* locked into fixed position and may be moved by the display software for

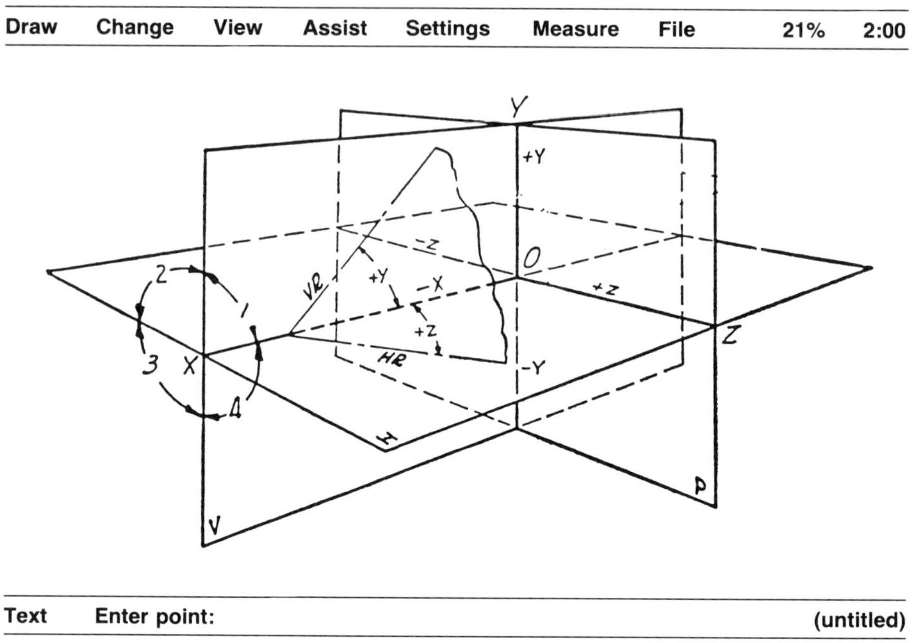

Figure 3.1 Coordinate model for plane surface *R*.

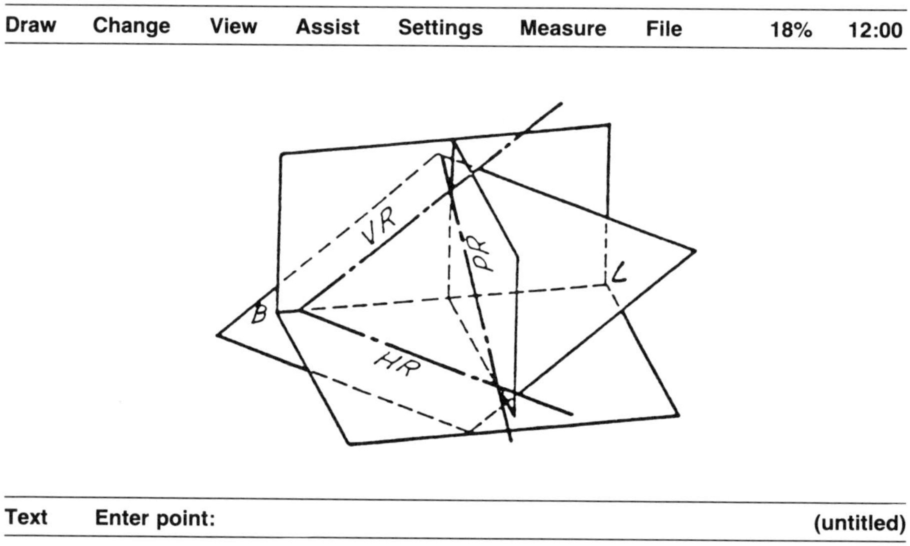

Figure 3.2 Display model for plane surface *R*.

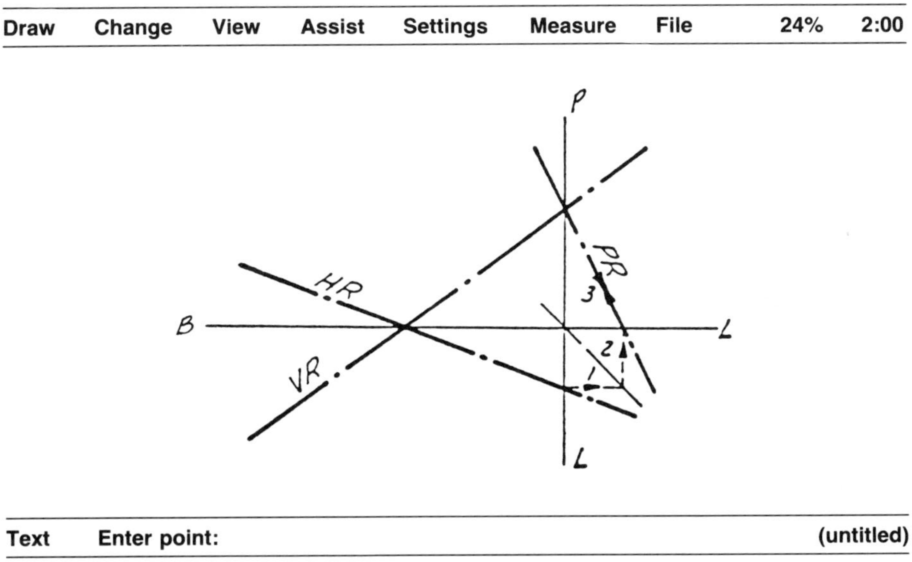

Figure 3.3 Typical screen display for plane surface *R*.

easy viewing. As shown in Figure 3.1, the coordinate planes intersect in *OX*, *OY*, and *OZ*. The origin *O* is the intersection of the three planes. Compare Figures 1.1 and 3.1; notice that the coordinate viewing system has been rotated about the *Y* in Figure 3.1 and that a plane surface labeled *R* is shown as trace *VR* and trace *HR*.

The positions and directions of the *H* and *V* traces of a plane are designated by coordinates and angles as follows:

> X = distance of the point of intersection of the traces with *BL*, to the right or left of the origin, as $+X$ or $-X$.
> Y = angle *V* trace *(VR)* makes with *H*, above or below *H*, as $+Y$ or $-Y$.
> Z = angle *H* trace *(HR)* makes with *V*, before or behind *V*, as $+Z$ or $-Z$.

Angles *Y* and *Z* are measured from the portion of *BL* to the right of the intersection of the traces.

Figures 3.2 and 3.3 represent the same plane surface *R*. In Figure 3.2 a model representing a convenient viewing location for *BL* is displayed. In Figure 3.3 the model has been revolved about *BL* so that *PL* can be labeled.

3.2 *H* AND *V* TRACE INTERSECTIONS

The plane *R* shown in Figure 3.1 is oblique to *BL* and intersects it at a point common to *V* and *H*. Likewise the plane is oblique to *PL* and intersects it at a point common to *V,P* and *H,P*. Three points of intersection are displayed in Figure 3.2., two points of intersection in Figure 3.3, and only one point in Figure 3.1. It is common to show only a single point of intersection for a plane. The most often displayed point of intersection is the intersection of *VR* with *HR* along *BL*. *PR* intersections are less common.

3.3 *P* TRACE INTERSECTIONS

The profile trace of a plane is its line of intersection with *P*. This trace cannot be parallel to the *V* trace unless the plane is perpendicular to *H*. This *P* trace cannot be parallel to the *H* trace unless the plane is perpendicular to *V*. The profile trace, therefore, intersects both the *H* trace and the *V* trace as shown in Figure 3.2. Since it is common to show the intersection of *V* and *H* as the base line *(BL)*, as in Figure 3.3, the profile plane is moved into prominence and labeled *PL*. The intersection of *VR* and *HR* (vertical and horizontal traces) remains fixed, however. Once an intersection has been established its value along a coordinate axis does not change, but its appearance on the display screen may change by revolution of the display model. As *H* and *P* are revolved into coincidence with *V*, Figure 3.3 is produced.

3.4 PLANES PARALLEL TO *H*

Figure 3.4 demonstrates a plane parallel to the *H* coordinate plane. Case 1 is the coordinate model display, and case 2 is the typical display screen showing *BL* and *PL* orientation for plane *W*. The trace is parallel to *BL*, and the *H* trace *(HW)* is at infinity. Therefore only the *VW* is displayed at the left of *PL*, and the *PW* is displayed at the right of *PL* in case 2.

3.5 PLANES PARALLEL TO *P*

Figure 3.5 demonstrates a plane parallel to the *P* coordinate plane. Case 1 is the model display while case 2 is technical descriptive geometry display for plane *R*. The trace is parallel to *PL*, and the *P* trace *(PR)* is at infinity. Therefore only the

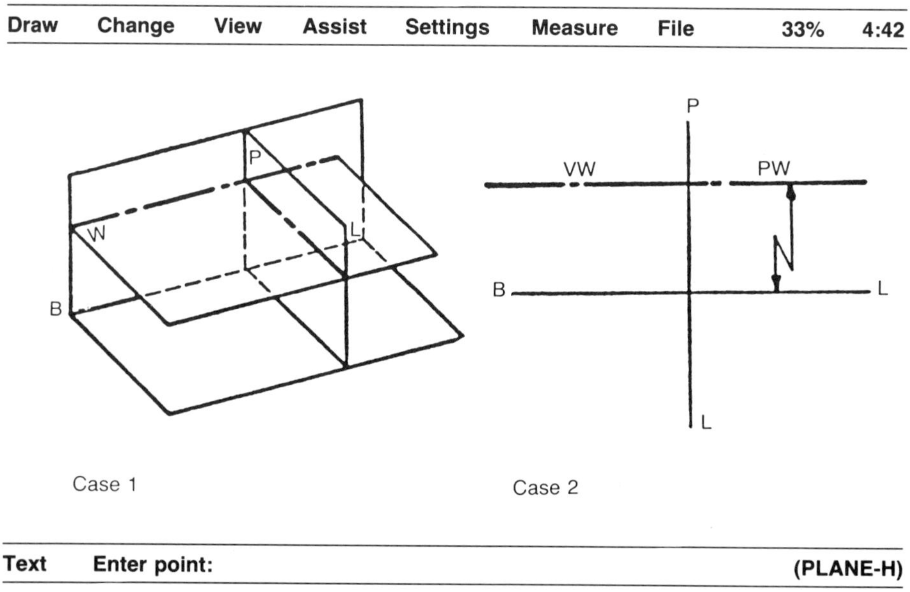

Figure 3.4 Plane parallel to *H*.

VR and *HR* traces can be displayed perpendicular above and below *PL* as shown in case 2.

3.6 PLANES PARALLEL TO *V*

Figure 3.6 demonstrates plane *U* parallel to the *V* coordinate plane. Case 1 is the model displaying the relationship, and case 2 shows the location for *HU* (horizontal

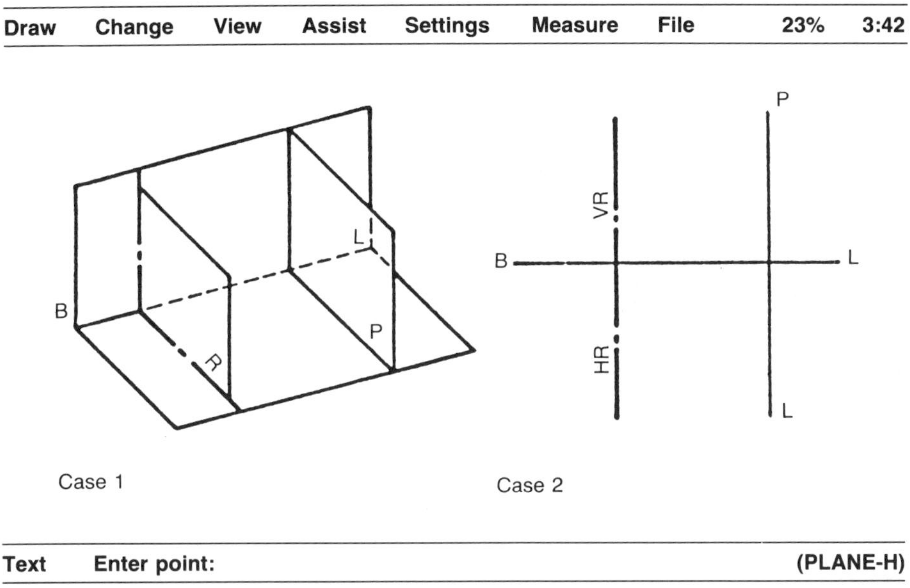

Figure 3.5 Plane parallel to *P*.

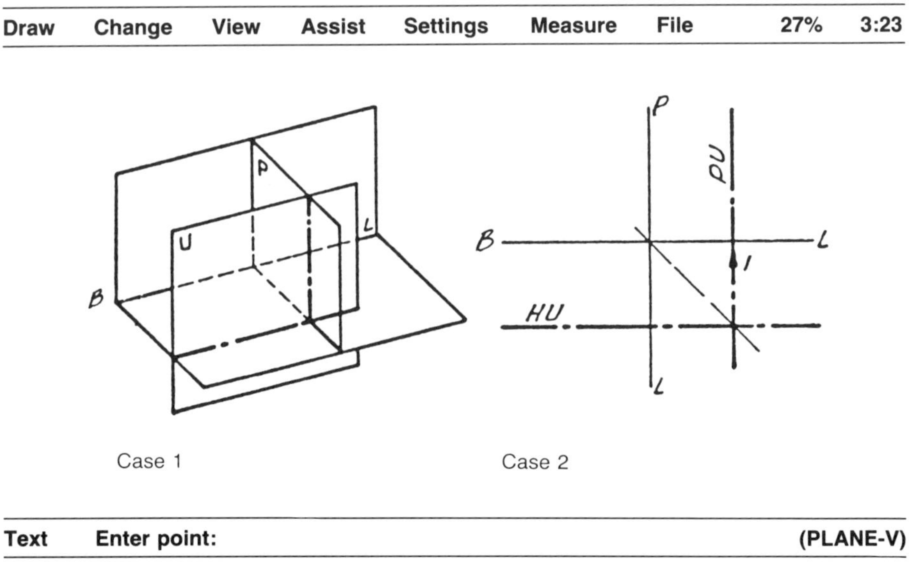

Figure 3.6 Plane parallel to *V*.

trace) below and parallel to *BL*. Case 2 indicates the location for *PU* (profile trace), since the *V* trace is at infinity.

3.7 PLANES PERPENDICULAR TO *H* AND INCLINED TO *V*

Case 1 of Figure 3.7 represents a plane surface called *S* that is perpendicular to *H* (*V* trace perpendicular to *BL)* and inclined to *V (H* trace inclined to *BL*). Case 2 clearly depicts the perpendicular relationship of *VS* and *HS*, while *PS* is perpendicular to *HS* and parallel to *VS*.

3.8 PLANES PERPENDICULAR TO *V* AND INCLINED TO *H*

Figure 3.8 (case 1) illustrates a plane surface called *T* that is perpendicular to *V* and inclined to both *H* and *P*. Case 2 clearly shows the angle of inclination called *slope*. Here the *VT* representation of the plane trace intersects the *H* plane and forms the slope of the plane surface *T*. Notice that the *HT* and *PT* are parallel to *PL* and *BL*.

3.9 PLANE PARALLEL TO *BL* AND INCLINED TO *H* AND *V*

Case 1 of Figure 3.9 indicates that plane surface *X* is inclined to both *H* and *V*. The angle of plane surface *X (PX)* is known as the *dip* of the plane. Case 2 shows the two-step method for determining the dip of the plane surface *X*.

3.10 PLANE INCLINED TO *H, V* AND *BL*

Case 1 of Figure 3.10 shows the *strike* of plane surface *Y*. The strike of the plane is read along *PY* shown in case 2. Whenever a plane is inclined to all three coordinate axes within a geometric model like that shown in Figure 3.10, a strike notation can be used.

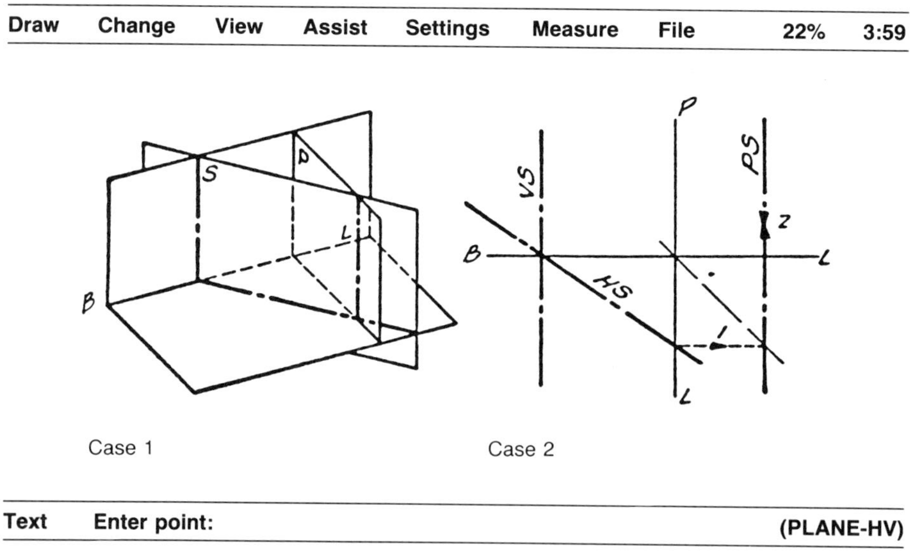

Figure 3.7 Plane perpendicular to *H* and inclined to *V*.

3.11 SINGLE LINE NOTATION FOR PLANES

The trace of a plane surface can be shown in a geometric model as indicated throughout this chapter. Plane surfaces can also be defined by the slope of the plane (Fig. 3.8), dip of the plane (Fig. 3.9), or the strike of the plane (Fig. 3.10). Often a plane is defined by a series of single line notations such as the trace and slope

Draw Change View Assist Settings Measure File 22% 3:59

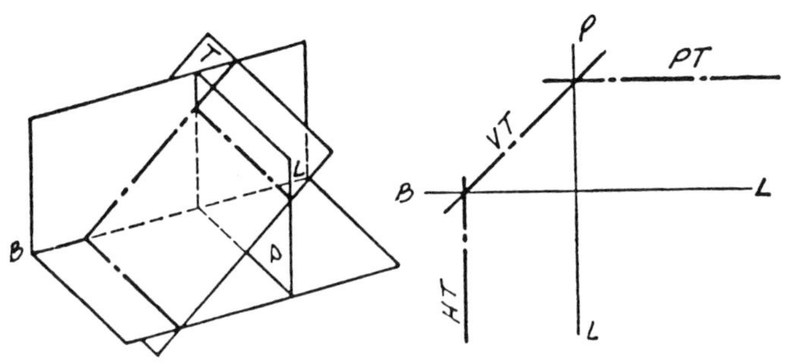

Case 1 Case 2

Text Enter point: (PLANE-HV)

Figure 3.8 Plane perpendicular to *V* and inclined to *H*.

Draw Change View Assist Settings Measure File 28% 2:45

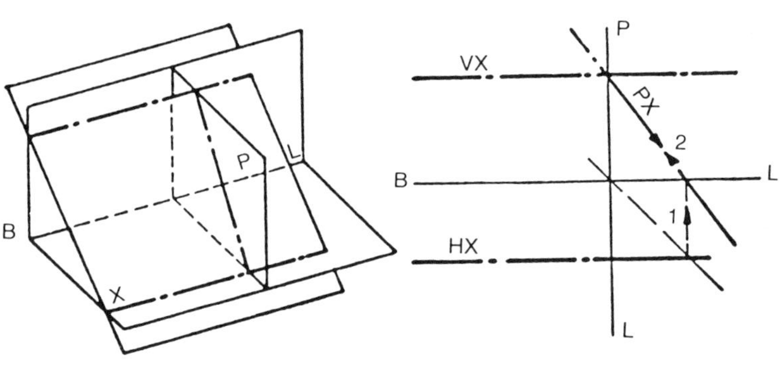

Case 1 Case 2

Text Enter point: (PLANE-BL)

Figure 3.9 Plane parallel to *BL* and inclined to *H* and *V*.

Draw Change View Assist Settings Measure File 22% 1:45

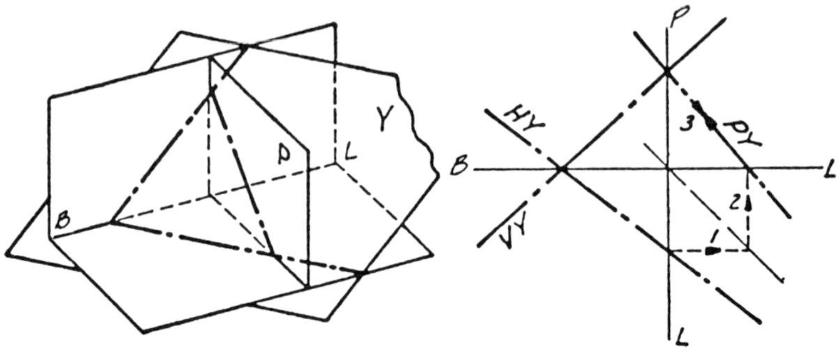

Case 1 Case 2

Text Enter point: (PLANE-PL)

Figure 3.10 Plane inclined to *H, V,* and *BL*.

or trace and dip. The coordinate models shown in case 1 situations of Figures 3.4 through 3.10 indicated the pictorial relationship of plane surface to reference surfaces. Case 2 portions of these figures depicted the geometric relationship and demonstrated single line notation for planes. Whenever a single line defined a complete plane surface, it contained one or two arrowheads, indicating that the trace of the plane can also be measured as to slope, dip, or strike.

3.12 CHAPTER PROBLEMS

The problems appearing at the end of this chapter have been selected to introduce the basic type of plane surface problems encountered by an engineer or engineering technician using a CADD display system. These problems emphasize the interdisciplinary aspects of geometric construction based upon a coordinate model introduced in Chapter 1 of this text.

PROBLEM SET 1 *Plane orientation*

In each problem display the plane defined.

1.1 *R* (Fig. 3.1)
1.2 *R* (Fig. 3.2)
1.3 *W* (Fig. 3.4)
1.4 *P* (Fig. 3.4)
1.5 *P* (Fig. 3.5)
1.6 *U* (Fig. 3.6)

1.7 *S* (Fig. 3.7)
1.8 *T* (Fig. 3.8)
1.9 *X* (Fig. 3.9)
1.10 *Y* (Fig. 3.10)

PROBLEM SET 2 *Plane definition*

In each problem display the type of plane called for.

2.1 Coord. (Fig. 3.1)
2.2 Plane (Fig. 3.2)
2.3 Oblique (Fig. 3.2)
2.4 Trace (Fig. 3.3)
2.5 Parallel (Fig. 3.4)
2.6 Principal (Fig. 3.5)
2.7 Profile (Fig. 3.6)
2.8 Perpendicular (Fig. 3.7)
2.9 Inclined (Fig. 3.8)
2.10 Parallel and inclined (Fig. 3.9)

4 Solid Definition

A solid is defined by boundary planes and is completely described by the projections of these planes on the coordinate model shown in Chapter 1, Figures 1.1 through 1.3. All planes on the boundary of a solid may be *coplanar, non-coplanar,* or a combination of both. For example, a simple six-sided cube contains only coplanar elements, a geodesic dome contains only non-coplanar elements, and a door stop contains both. Most CAD workstations are capable of displaying solids in a variety of options.

The options available for Autodesk, Inc., software presentation include

1. With a wireform in any axonometric orientation (Vpoint command), shaded polygon surfaces can be added to simulate halftone presentations (3-D commands).
2. Solid models with hidden line removal in any axonometric rotation (HIDE and SOLID commands), and
3. Light source for realistic model shades and shadows to include basic primitives for cube, cone, pyramid, sphere, and torus.

With these building blocks, realistic models having surprising detail as shown in Figure 1.6 can be displayed. AUTOSOLID and AUTOSHADE software were selected because they were already in assembly language format and part of the AutoCAD software package used throughout the other chapters of this book.

4.1 CONCEPTS OF SOLID DEFINITION

This chapter also contains the use of solids as defined within the Autodesk, Inc., software AUTOSOLID. This denotes theoretical solids and has the same meaning as in ANSI Y14.5M; therefore solid descriptions will represent a basis for an object from which permissible variations may be established by augmenting physical characteristics. Geometric descriptions include solid building blocks that consist of points, lines, curves, and surfaces to define the solid. Topological descriptions cover the pertinent connectivity relationships for the solid. The information structure within AUTOSOLID and explained in this chapter is a language format that anticipates the need to associate nongeometric data with various levels of object shape data for solids.

4.2 INFORMATION STRUCTURES FOR SOLIDS

The information structure within AUTOSOLID is a machine- and human-readable language with a well-defined syntax. The unit of information employed here may be (1) Domain, (2) Geometric, (3) Topological, or (4) Miscellaneous.

A Domain structure defines a set of solid data in terms of its members, which are a finite collection of numerical values or character strings representing attributes. This structure is particularly useful in terms of extensibility into the areas of product definition data for CAD/CAE.

The Autodesk, Inc., Geometric structure describes geometric entities (point, line, plane, or solid combinations) used to construct a solid. Methods used to construct solids are interpolation, generation, point-set operations, replications, and a small selection of special operations.

Topological structures define relationships among vertices, edges, faces, and objects in terms of their interior boundaries. These structures are built from simple entities to more complex.

4.3 SOLID ENTITIES

Three classes of entities are used to create Autosolids. These are Geometric, Topological, and Miscellaneous. Geometric and Topological have just been defined; Miscellaneous entities are composed of only the *group* entity, which is a collection of other entities (smaller solid subparts). A collection of entities described through the same structure type is known as a *relation*. A relation is communicated as shown in Chapter 3 for a plane surface.

A plane represents a relation for a linear blend (interpolation) of points that come from specified domains. This data may be expressed in a language format as

PLANE/LIN/ENT0, ENT1, PNTA, PNTB:P1(P1, P2, P3)

The parameter data for a plane may be shown as

Parameter	Value	Format	Comment
1	n	I	Number of points in plane
2	ie	I	Pointers used in plane surface to show order of connections
3	m	I	Number of properties for plane
4	rel	A	Language format used

The parameter data for a solid may be shown as

Parameter	Value	Format	Comment
1	n	I	Number of lines in the boundary
2	ie	I	Pointers used in solid edges to show order of connections
3	m	I	Number of properties (surfaces) for the solid
4	rel	A	Language format used (AutoLISP)

A linear view of the parameter data for a plane is

AutoList	**DIR Entry**
Lines	Rel entity, name & pointer

PLANE/LIN/ENTO, ENT1/POINTS, POINTS:
P1(P1,P2,P3);

A linear view of the parameter data for a solid is

AutoList	**DIR Entry**
Edges	Rel entity, name & pointer

LINES/LIN/ENTO, ENT1/POINTS, POINTS:
S1(P1, P2); L2(P2, P3) ....LN(Pn, Pn + 1)

Figure 4.1 shows the Autodesk, Inc., display for a solid (S1) as a linear blend (interpolation) of points that come from specified domains. A better understanding of communication of language format via file structure can be seen from an Autodesk, Inc., example shown in Figure 4.2. The edges of a solid are displayed in a connected format. This type of screen display is called the wireform of the solid.

The completed solid statements for Figure 4.2 are shown displayed in Figure 4.3, and a portion of the relational table format is shown in Figure 4.4. A file

Figure 4.1 Structure and its components. (Courtesy Autodesk, Inc.)

structure portion is shown in Figure 4.5, and a sample portion of the linear file format is shown in Figure 4.6.

4.4 SOLID GENERATION

Where solid generation is used, it is generally explained in context. However, the software generation provided in Figure 4.7 is used throughout this chapter.

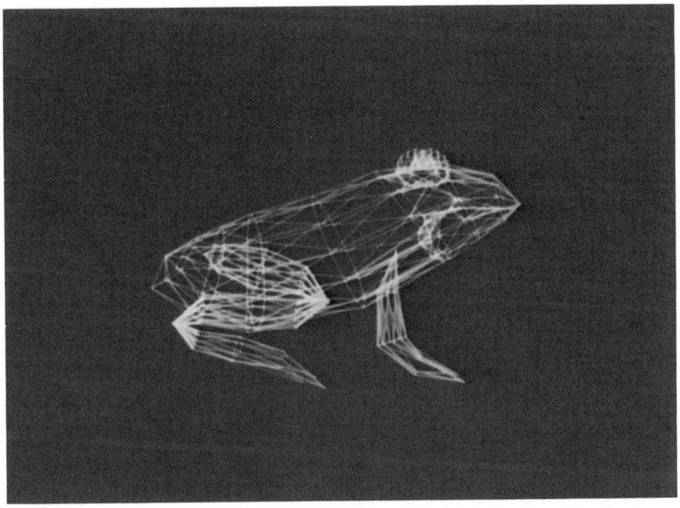

Figure 4.2 Wireform of a leaping frog.

Figure 4.3 Shaded surfaces for leaping frog.

	LFROG		PNTR001
POINTS	X	Y	Z
P1	0.	1.	0.
P2	0.	2.	.185
P3	.185	0.	0.
P4	1.	2.250	0.
.			
.			
.			
.			
.			
.			
.			
P77	0.	0.	7.500

PNTR001: '77 POINTS'

	LFROG	ENTR001
LINES	ENTO	ENT1
G1	P1	P2
G2	P2	P3
G3	P3	P4
.		
.		
.		
.		
.		
.		
.		
.		
G139	P47	P77

ENTR001: '139 LINES'

Figure 4.4 Relational table format for leaping frog.

4.5 GEOMETRIC STRUCTURES

The Geometric structure describes the shape and mathematical representation of the defined solid. The solids specified in this chapter are divided into six classes: (1) Primitive, (2) Interpolative, (3) Generative, (4) Special, (5) Point set, and (6) Replicative solids. These solids are displayed constructively with the Primitive structure being the basic building block.

```
LFROG FILE LISTING

LFROG,POINTS'77 POINTS'/PNT/X,Y,Z/3REAL: P1(0.,1.,0.),

PT2(0.,2.,.185), PT3(.185,0.,0.), PT4 (1.,2.250,0.), PT5

.................................................................

.................................................................

PT77(0.,0.,7.500);

LINES/LIN/ENTO,ENT1/2*POINTS:G1(P1,P2),G2(P2,P3),G3(P3,

P4),.............................................................

.................................................................

G139(P47,P77);

PTYPE/DOM/5*CHAR,CLOS:'DOT','CROSS';

POINTR/PNT/X,Y,Z,KIND/3*REAL,PTYPE:CPT(0.,0.,5.'CROSS');

LTYPE/DOM/7*CHAR,OPEN:'DASHED','PHANTOM';

     CURVES/CSTR/N,N*ENT,(N-1)*PAR/INTR(1,*);
```

Figure 4.5 File structure for leaping frog.

The Primitive, Interpolative, Generative and Special solids are displayed as parametric entities; that is, the entities are displayed as images of $I(n)$, for $n = 0,1,2$ or 3, under a continuous function. $I(n)$ is called the parameter space, and the parameters are named u,v,w; Table 4.1 relates these concepts to dimension and entity type.

Table 4.2 lists all possible geometric structures. The Point set and Replicative structures are not parametric but are displayed as transformations represented by entities.

4.6 PRIMITIVE STRUCTURES

From Table 4.2 we note that point and marker, shown in Figure 4.8, comprise the only two such structures. Autodesk, Inc., uses a marker system called UCS (user coordinate system) which defines the precise location of solid geometry in space. The origin of a UCS can be placed anywhere within the display area, and the UCS axes can be rotated or tilted at any angle as shown in Figure 4.9. The locations of

LFROG FILE LISTING

ENTITY	DIR ENTRY
77 POINTS	rel entity, name & pointer
139 LINES	pointer to DE
43 SHADED SURF	associativities & properties
PTYPES	domain structures
POINTR	relation entity
LTYPES	pointer to PD
CURVES	language format

Figure 4.6 Linear file for leaping frog.

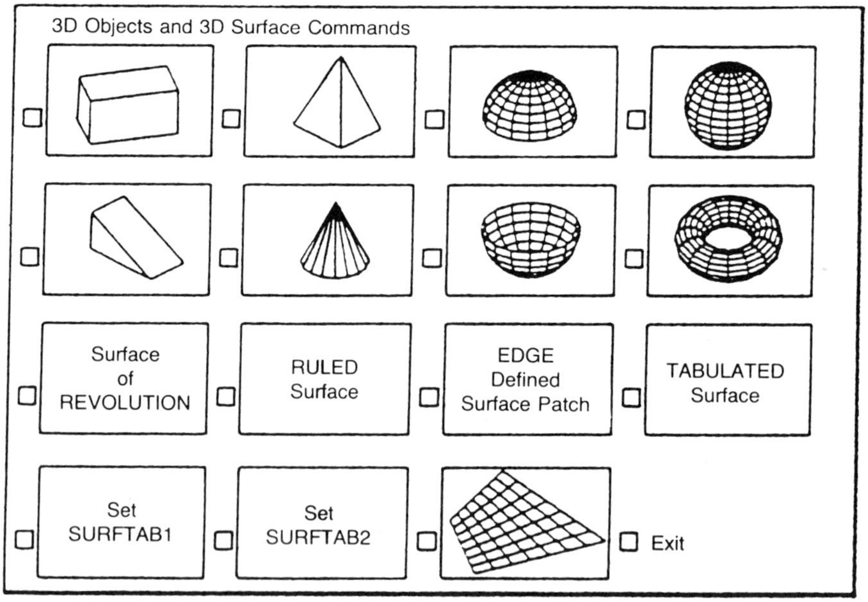

Figure 4.7 Solid generation from an Autodesk, Inc., software menu.

Table 4.1 DIMENSIONALITY AND PARAMETERIZATION

Structure	Dimension	Space	Parameter
POINT	0	0	0
LINES	1	I	u
SURFACES	2	I(2)	u, v
SOLIDS	3	I(3)	u, v, w

Table 4.2 GEOMETRIC STRUCTURES

Structures	Solid	Surface	Line	Point
Primitives				
Point	—	—	—	√
Marker	—	—	√	—
Interpolative				
Line	—	—	√	—
Arc	—	—	√	—
Cubic	√	√	√	—
Nth-degree poly	√	√	√	—
Spline	√	√	√	—
Generative				
Translation	√	√	√	—
Rotation	√	√	√	—
Special				
Reverse	—	—	√	—
String	—	—	√	—
Flip	—	√	—	—
Point set				
Closure	√	√	√	√
Union	√	√	√	√
Intersection	√	√	√	√
Difference	√	√	√	√
Replicative				
Translation	√	√	√	√
Rotation	√	√	√	√
Scale	√	√	√	√
Mirror	√	√	√	√

the viewing axes are shown by the display of a small coordinate system icon shown in the lower left hand corner of the screen.

4.7 PRIMITIVE SYMBOLS

The UCS icon is shown in Figure 4.9 as it would be seen from different view points. A *W* displayed in the *Y* arm indicates that the current UCS coincides with the World Coordinate System. If a + appears at the lower left of the icon, it

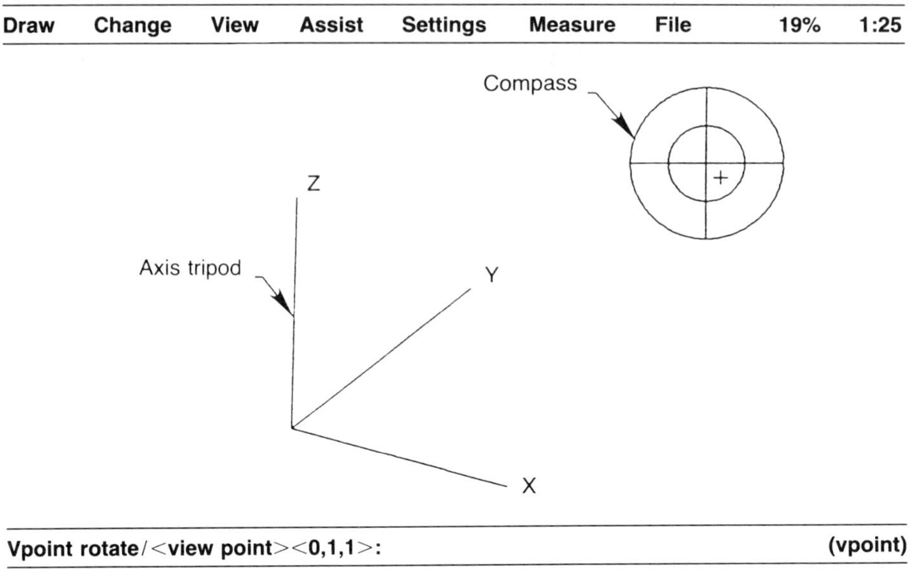

Figure 4.8 Vpoint command and compass system.

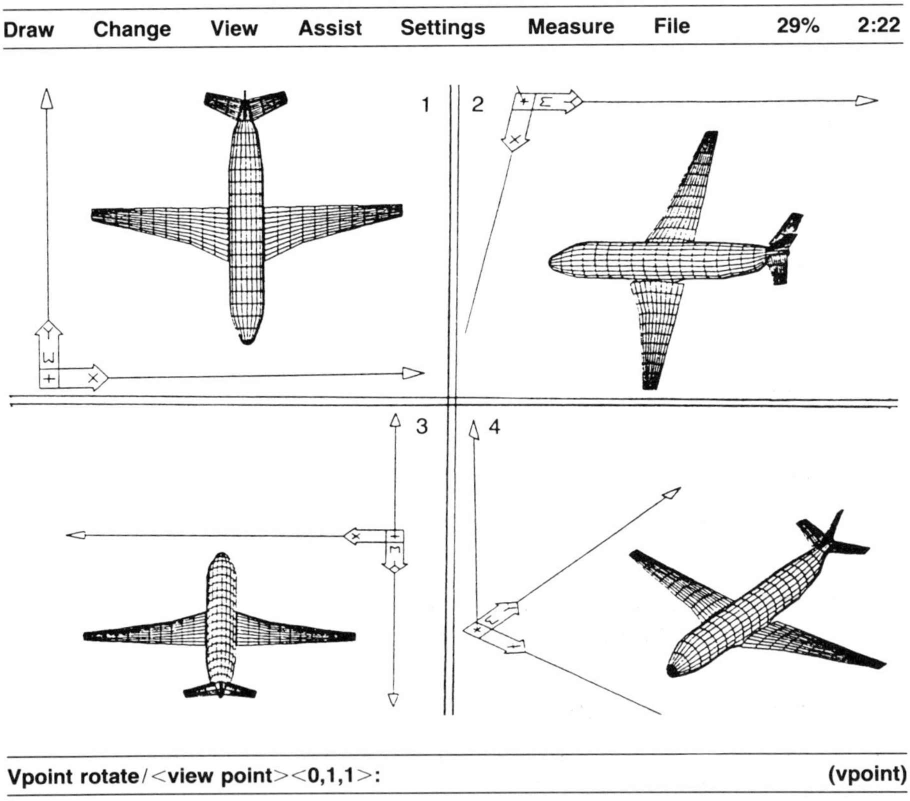

Figure 4.9 Vpoint command and VPORT system.

indicates that the icon itself is located at the UCS origin point. Autodesk, Inc., software may also display the UCS icon as a special symbol. When the *XY* plane of the current coordinate system is perpendicular to your viewing plane, a broken pencil inside a box is displayed, which indicates that selecting and drawing solids are limited in such a situation.

The icon is positioned by the Vpoint command. Refer to Figure 4.9 view port 3 to see the result of Vpoint 0,1,1; view port 2 contains 1, − 1,1; view port 1 contains 0,0,0. The Vpoint command also provides an option allowing you to specify a three-dimensional view point in terms of two angles: one within the *X-Y* plane of the current UCS measured counterclockwise from the *X* axis and the other one measured up from the *X-Y* plane. View port 4 contains the result of

Command: **Vpoint**
Rotate/<view point> <current>: ***R***
Enter angle in the *X-Y* plane from *X* axis <current>: **30**
Enter angle from the *X-Y* plane <current>: **50**

4.8 INTERPOLATIVE STRUCTURES

From Table 4.2 we see that Line, Arc, Cubic, Nth polynominal and Spline are all interpolative. Further study of display figures will indicate how Autodesk, Inc., statements are used to display each figure. For example, Figure 4.10 shows three examples of *line* interpolation. The line interpolation is used to create a 3-D polygon mesh of a surface entity defined by a matrix of 3-D vertices. You can specify as

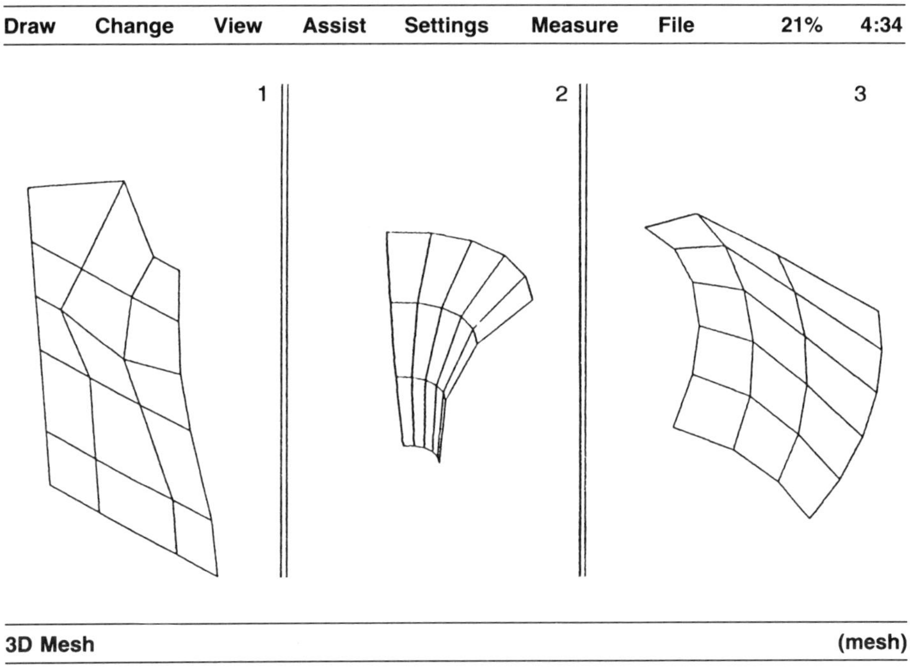

Figure 4.10 The 3DMESH command.

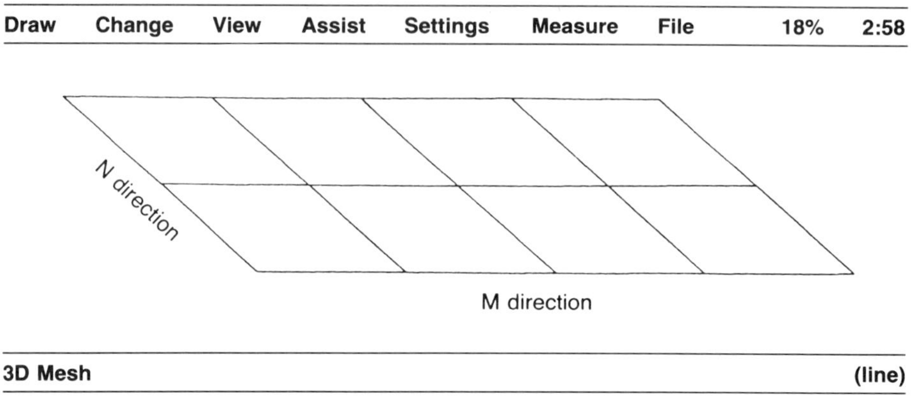

Figure 4.11 Linear interpolative display model.

many as 256 vertices in two directions and thus control the density or tabulations of the mesh to suit your solid application. The examples shown in Figure 4.10 consist of 24 vertices, 6 in one direction and 4 in the other direction.

3-D meshes are typically used to display partial sections of a solid model. These sections are then joined together to develop a complete model. There are several methods available in Autodesk, Inc., software to generate interpolative structures. To demonstrate each of these methods, a series of display figures will illustrate each method listed in Table 4.2 under *interpolative*. Figure 4.11 shows a linear interpolative display model. Using Autodesk, Inc., software, you can create a display figure by

1. Indicating the position of each vertex in the mesh using the 3DMESH command,
2. Creating edge geometry and interpolating a surface by using TABSURF, RULESURF, REVSURF, or EDGESURF commands, or
3. Placing control points and approximating a surface about the control points by using the following techniques:
   a. Quadratic approximation
   b. Cubic approximation
   c. Bezier approximation

This technique is demonstrated first by Figure 4.12 where a vertex by vertex mesh is defined in the *M* and *N* directions. This display was begun with a matrix of 3-D points. Although point entities are not needed to define a 3-D mesh, they are provided here to make the selection of each vertex in the mesh easier. You may display a similar mesh by following the prompts in this chapter section to select a "running" example called *NODE*.

COMMAND: **OSNAP**
OBJECT snap modes: **NODE**

When you enter the 3DMESH command, Autodesk, Inc., software will ask for the *M* and *N* size of the mesh. Let's say we want the *M* direction to be the

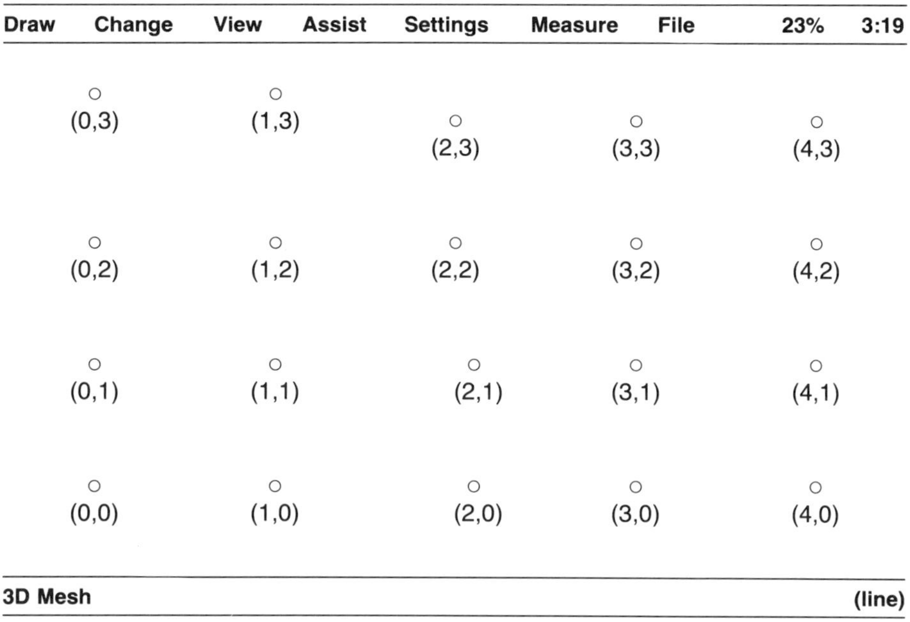

Figure 4.12 *M* and *N* point matrix display model.

horizontal direction and the *N* direction to be the vertical. Enter the following prompts:

COMMAND: **3DMESH**
Mesh *M* size: **5**
Mesh *N* size: **4**

So the *M* size is 5 and *N* size is 4. After the mesh size is specified, you may input vertex locations as

Vertex <0,0>: pick point 1 (shown in Figure 4.13)
Vertex <0,1>: pick point 2
: (Continue to pick points)
:
:
:
:
Vertex <4,3>: pick point 20

After you have picked each location as shown in view port 1 of Figure 4.13, a mesh is displayed for you. To get a better view of the mesh, view port 2 has been set to

COMMAND: **Vpoint**
Rotate/<View Point> <0.0,0.0,1.0>:
Enter angle in *X*-*Y* plane from *X* axis <270>:**315**
Enter angle from *X*-*Y* plane <0.,>:**15**

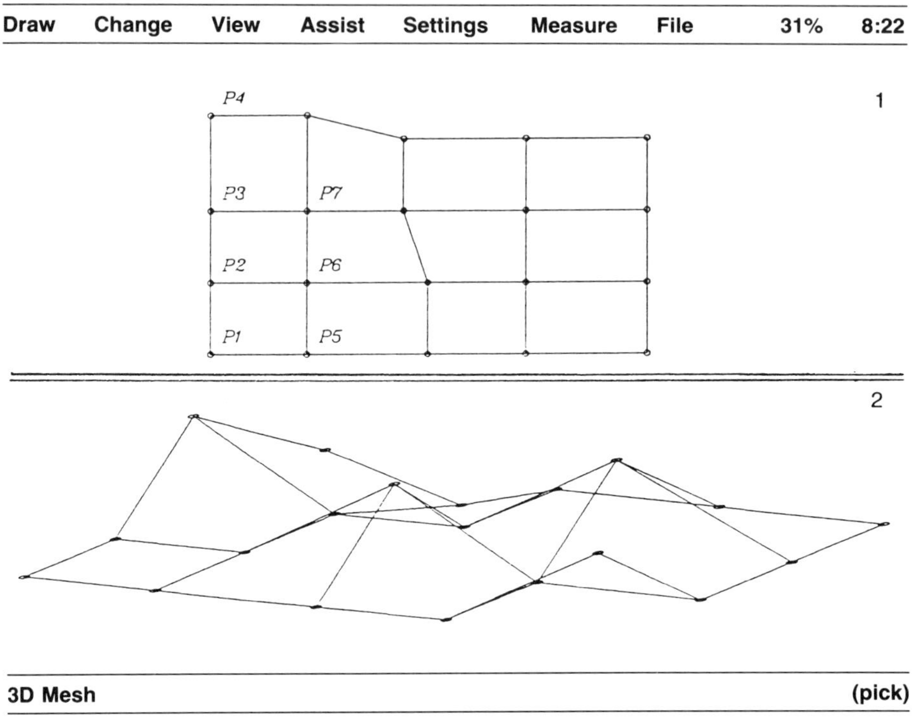

Figure 4.13 Vertex selection in a 3-D mesh display.

The display you have created can be moved, copied, rotated, or scaled just as any other display entity. In addition, the vertices within the mesh can be moved individually using the PEDIT command as

COMMAND: **PEDIT**
Select polyline: <pick mesh on display screen>
Edit vertex/Smooth surface/Desmooth/Mclose/Nclose/Undo/eXit/<X>:

Let's change the 3-D location of a vertex in the mesh as shown in Figure 4.14 view ports 1 through 4. Respond to the PEDIT prompt with **E** and an x is displayed as shown in view port 1. This is the vertex to change. View port 2 shows how the "left/right/up/down" options will be used.

View port 3 shows the pick box for point *1*, and view port 4 shows the results of

Vertex (0.0). Next/Previous/Left/Right/Up/Down/REgen/eXit<N>:**M**
Enter new location: **−2**

4.9 ARC INTERPOLATION

An arc defined mesh is a mesh displayed from existing interpolation geometry. For example, the curved surface in Figure 4.15 is interpolated between four previously created arcs. Although arc generated meshes have M and N directions just like line

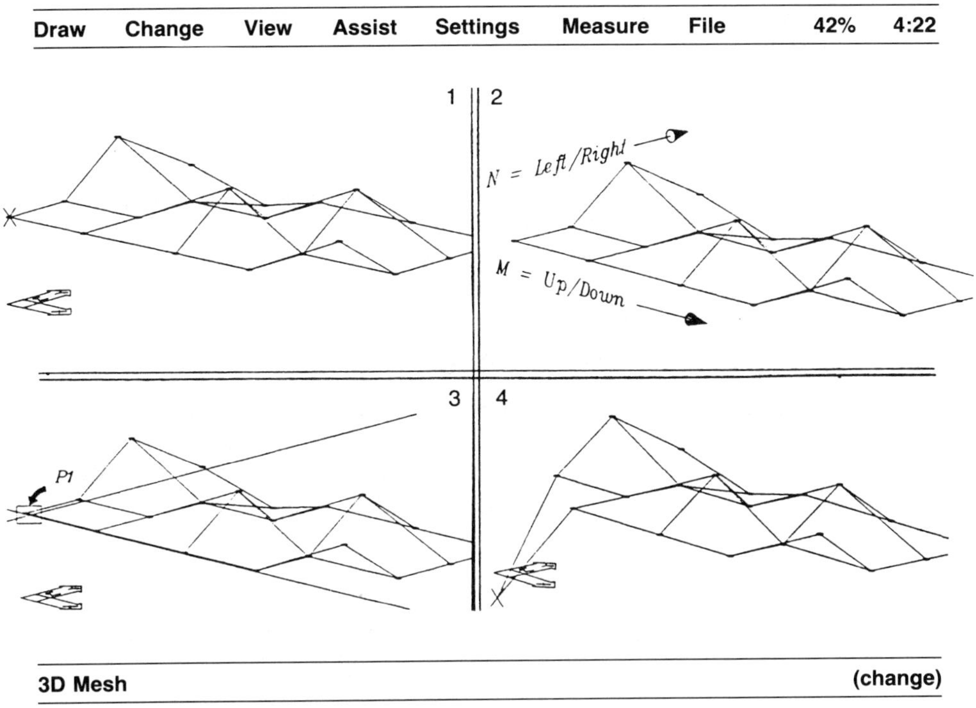

Figure 4.14 PEDIT for mesh changes.

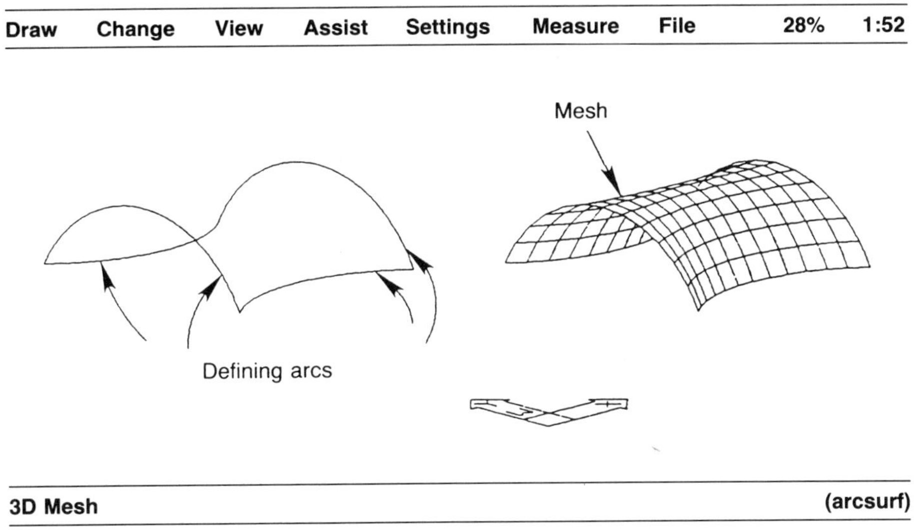

Figure 4.15 Using arc interpolation.

meshes, the mesh size is specified differently. There are two Autodesk, Inc., variables called SURFTAB1 and SURFTAB2 used to control the density.

4.10 CUBIC INTERPOLATION

A cubic defined solid is a solid displayed from existing interpolation geometry. For example, the curved solid in Figure 4.16 is interpolated between display points previously created from an equation. The Cubic method of interpolation is primarily intended to be used to create application specific solids with AutoLISP in a language-based format.

4.11 APPROXIMATE INTERPOLATIONS

A cubic defined solid is one of three approximations provided by Autodesk, Inc. Quadratic and Bezier are also available. Each type of approximation will display a slightly different surface for a solid as shown in Figure 4.17. The system variable SURFTYPE 5 will select Quadratic, 6 will select Cubic, and 8 will select Bezier.

4.12 GENERATIVE STRUCTURES

A generative structure requires the ability to translate and rotate images in the creation of the solid. Autodesk, Inc., provides two commands: REVSURF shown in Figure 4.18 and RULESURF shown in Figure 4.19.

The REVSURF command generates a surface of revolution by sweeping a profile path around a rotation axis. The path curve can be a 2-D or 3-D polyline, line, arc, or circle. The density of the surface mesh is controlled by setting SURFTAB1 *(M)* and SURFTAB2 *(N)* as

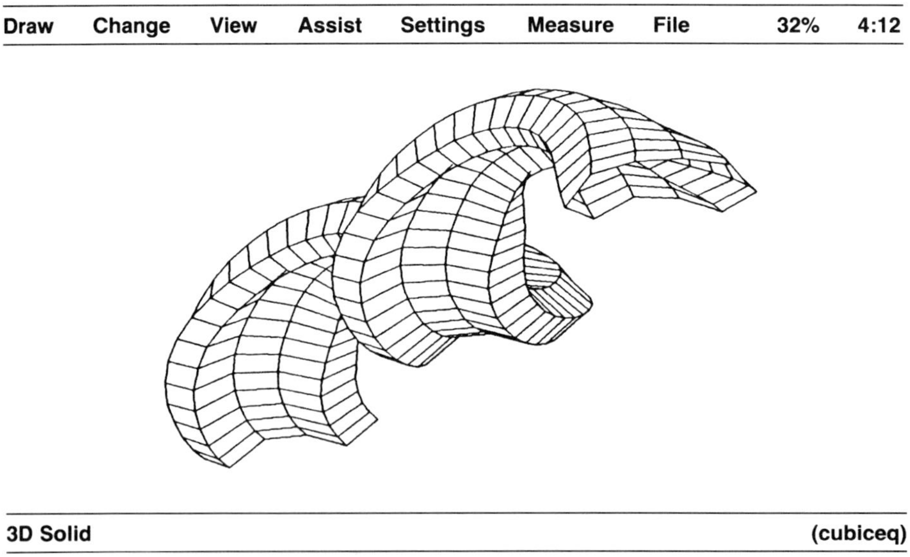

Figure 4.16 Using cubic interpolation.

COMMAND: **REVSURF**
Select path curve:
Select axis of revolution:
Start angle<180.>:
Included angle(+ = ccw, − = cw)<450.>:

The RULESURF command generates a surface of straight tabulation lines translated along the path curves, with the end points of the tabulation lines touching

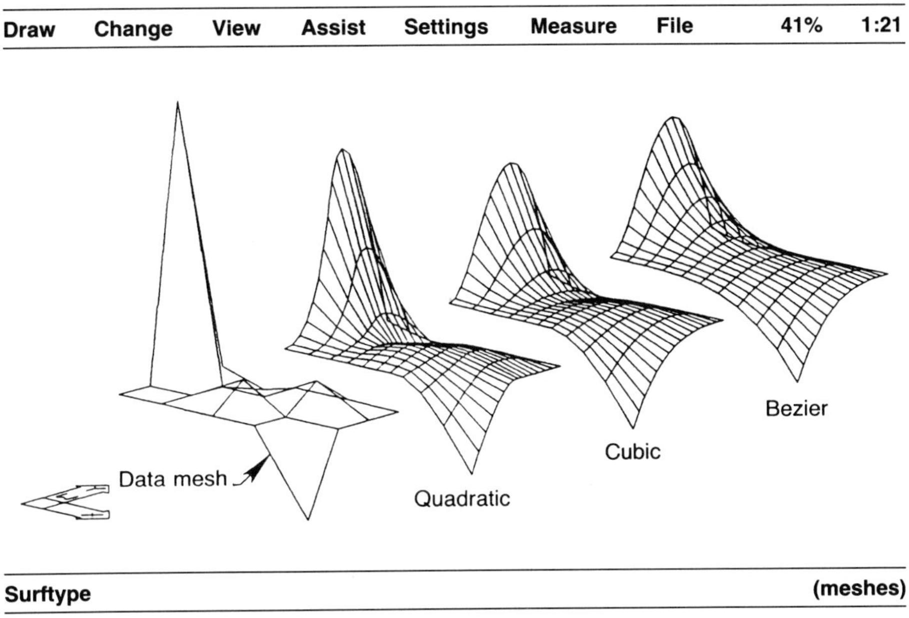

Figure 4.17 Using cubic, Bezier, and quadratic interpolation.

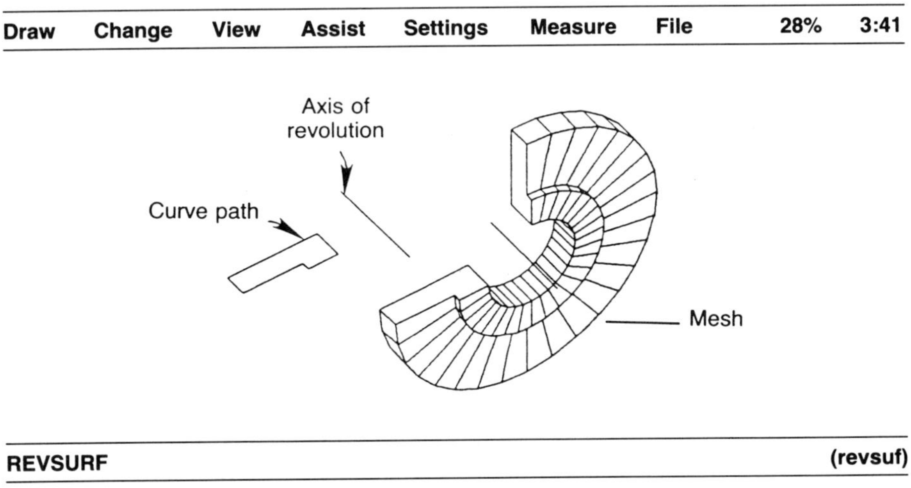

Figure 4.18 Solid from REVSURF command.

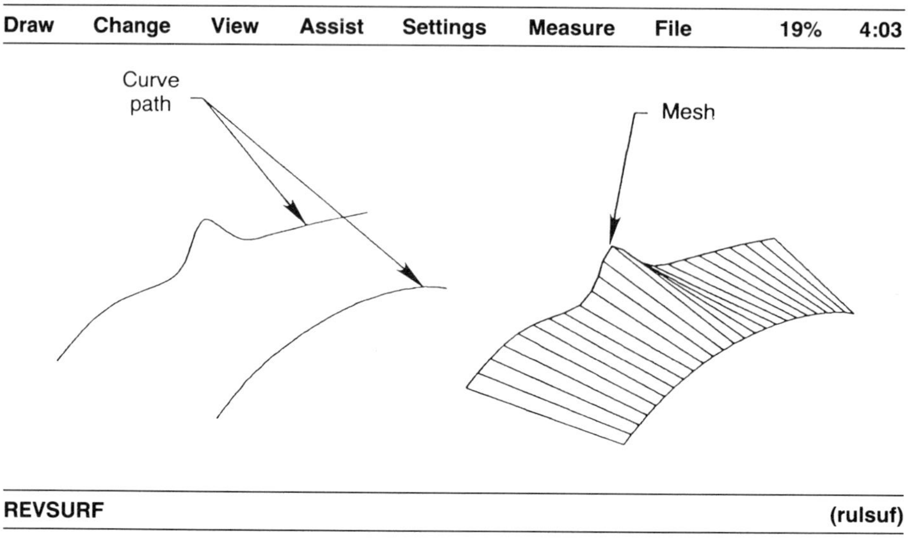

Figure 4.19 Surface from RULESURF command.

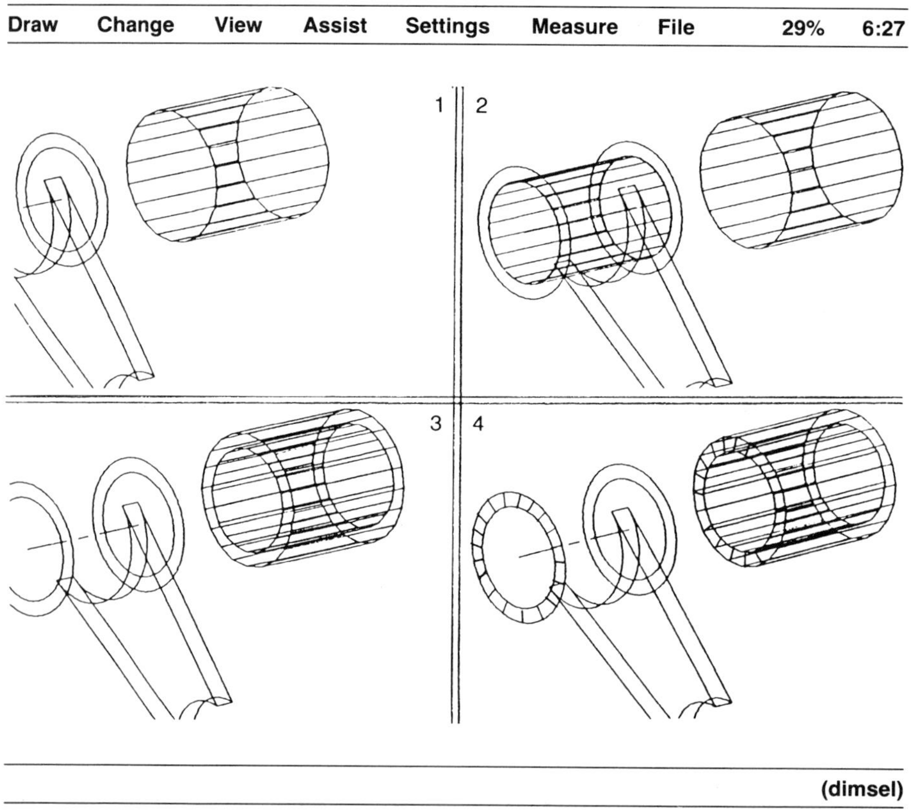

Figure 4.20 Dimension-selecting closure examples.

the path curves at specified intervals. The surface swept out by the lines creates a ruled surface mesh for the solid as

COMMAND: **RULESURF**
Select first defining curve:
Select second defining curve:

4.13 SPECIAL STRUCTURES

From Table 4.2 we note that four types of structures are used to orient solids. These are not used to display or build solids but are used to view solids. Autodesk, Inc., provides the DVIEW command for this purpose. This allows the dynamic previewing of parallel or perspective views of solids through the use of "slider bars." This command uses a *camera* and *target* to relate the DVIEW options to the solids.

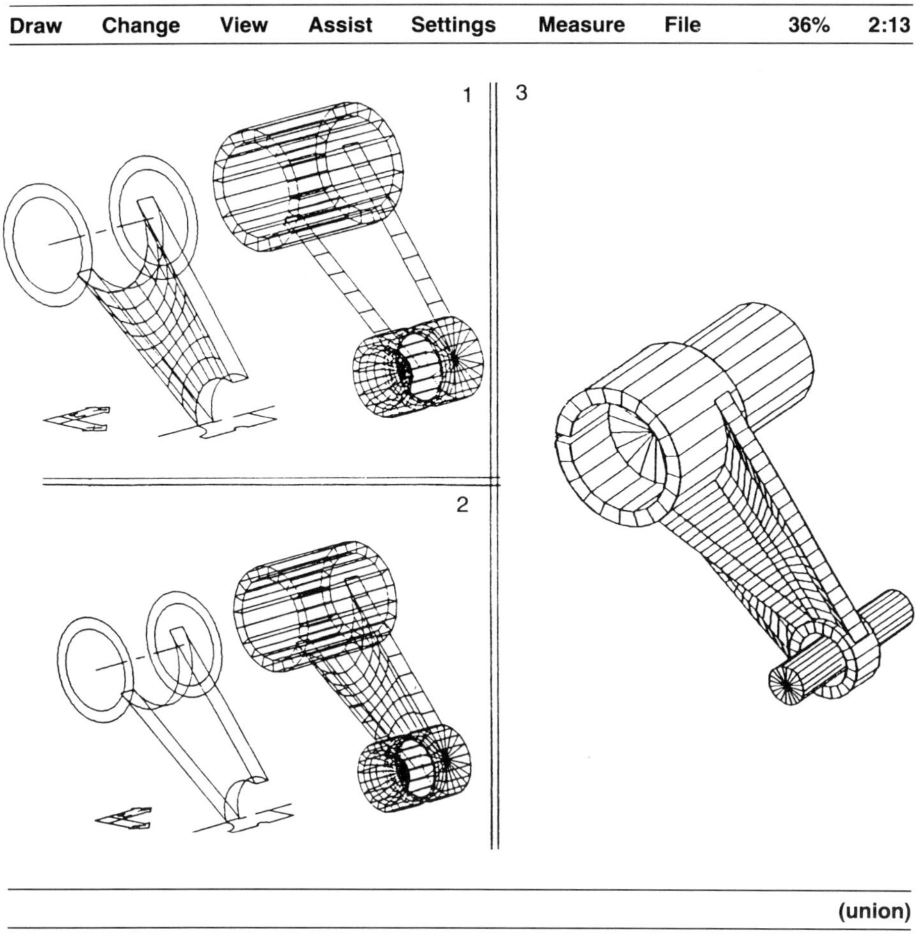

Figure 4.21 Union example.

You may practice the use of DVIEW with any saved solid object that you have created. The object of this example is to view the solid from different 3-D view points to get a better idea of how a solid is constructed. We will use DVIEW to specify target and camera points, change the view from the ground level to above the solid object, and walk around the solid dynamically.

Enter the DVIEW command and window the entire solid object when prompted:

COMMAND: **DVIEW**
Select objects: **w**
Other corner:
CAmera/TArget/Distance/POints/PAn/Zoom/TWist/Hide/Undo/<eXit>:**PO**
Enter target point<5′ − 5″,10′ − 0″,0′ − 0″>:
Enter camera point<5′ − 5″,10′ − 0″,0′ − 0″>:

The CA (camera) option allows you to dynamically rotate the camera point around the target. The CA point can be specified in terms of two angles, similar to the Vpoint command's ROTATE option. The first angle rotates from the *X-Y*

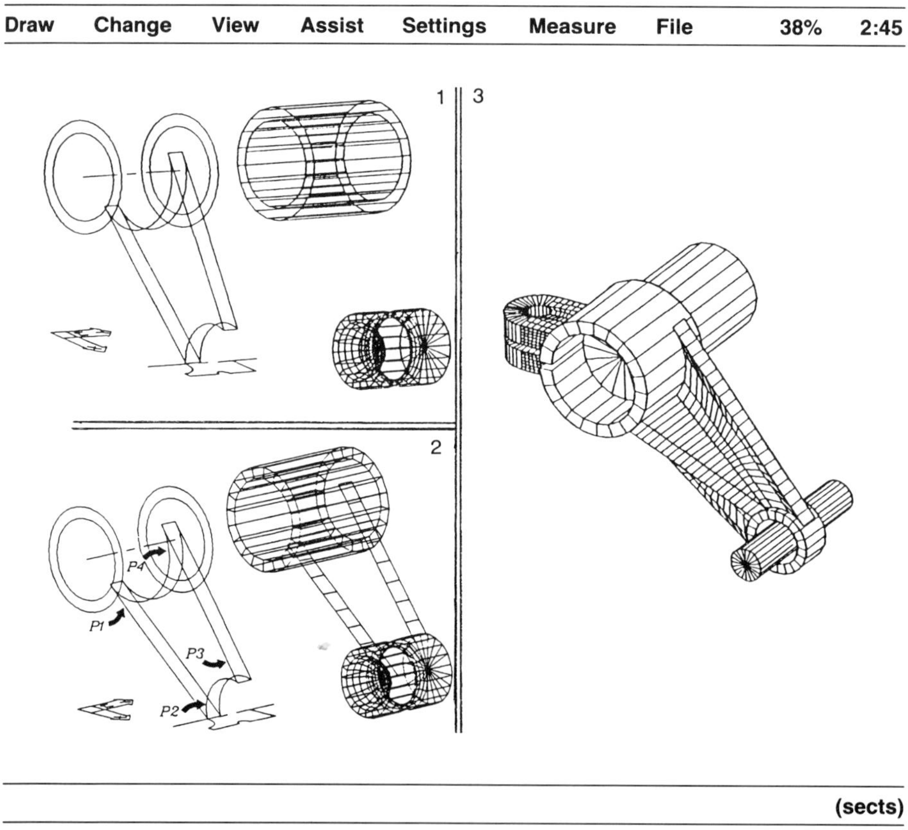

Figure 4.22 Intersection example.

plane, and the second rotates around the target point. The CA also allows you to preview what the resulting view will look like as you choose the two angles. When you enter the CA option, Autodesk, Inc., software displays a slider bar at the right of the screen. As you move the cursor within the slider bar, all of the objects you selected when you entered the DVIEW command will be visible as you rotate the camera around the target point. After the first angle is selected, another slider bar

Table 4.3 STRUCTURE AND SENSE PRESERVATION

Replication	Preserves Structure	Preserves Sense
TRANSLATION/ROTATION	Yes	Yes
ROTATION	Yes	Yes
TRANSLATION	Yes	Yes
SCALE	No	Yes
REFLECTION	Yes	No

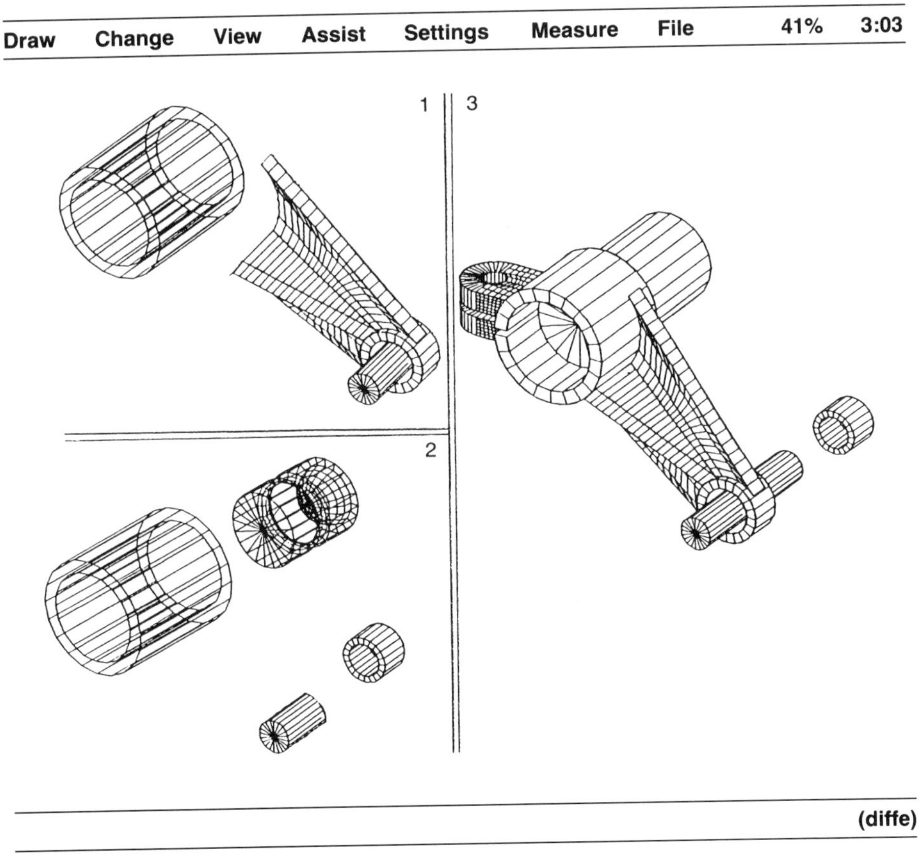

Figure 4.23 Difference example.

Table 4.4 REPLICATION STRUCTURES

Replication	T_1	R_1	S	R_2	T_2	Input Triples						
						Q1	Q2	Q3	R1	R2	R3	K
TRANSLATION/ ROTATION	$\mathcal{T}_{-A_1}$	$\mathcal{R}_\alpha^T$	I_4	$\mathcal{R}_\beta$	$\mathcal{T}_{B_1}$	√	√	√	√	√	√	X
ROTATION	$\mathcal{T}_{-A_1}$	$\mathcal{R}_\alpha^T$	I_4	$\mathcal{R}_\beta$	$\mathcal{T}_{A_1}$	√	√	√	= Q1	= Q2	√	X
TRANSLATION	$\mathcal{T}_{-A_1}$	I_4	I_4	I_4	$\mathcal{T}_{B_1}$	√	X	X	√	X	X	X
SCALE	$\mathcal{T}_{-A_1}$	I_4	$\begin{bmatrix} KX & & & 0 \\ & KY & & \\ & & KZ & \\ 0 & & & 1 \end{bmatrix}$	I_4	$\mathcal{T}_{A_1}$	√	X	X	X	X	X	√
REFLECTION	$\mathcal{T}_{-A_1}$	$\mathcal{R}_\alpha^T$	$\begin{bmatrix} 1 & & & 0 \\ & -1 & & \\ & & 1 & \\ 0 & & & 1 \end{bmatrix}$	$\mathcal{R}_\alpha$	$\mathcal{T}_{A_1}$	√	√	√	X	X	X	X

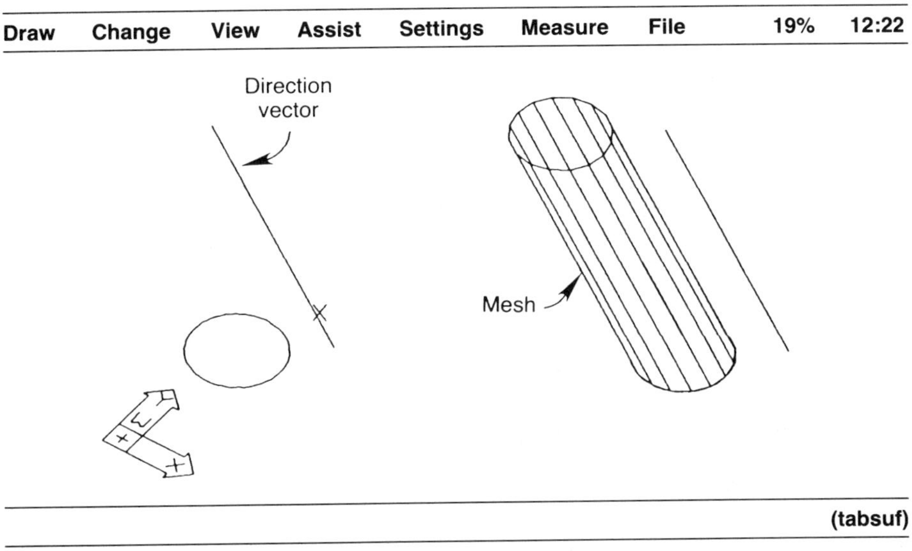

Figure 4.24 TABSURF example.

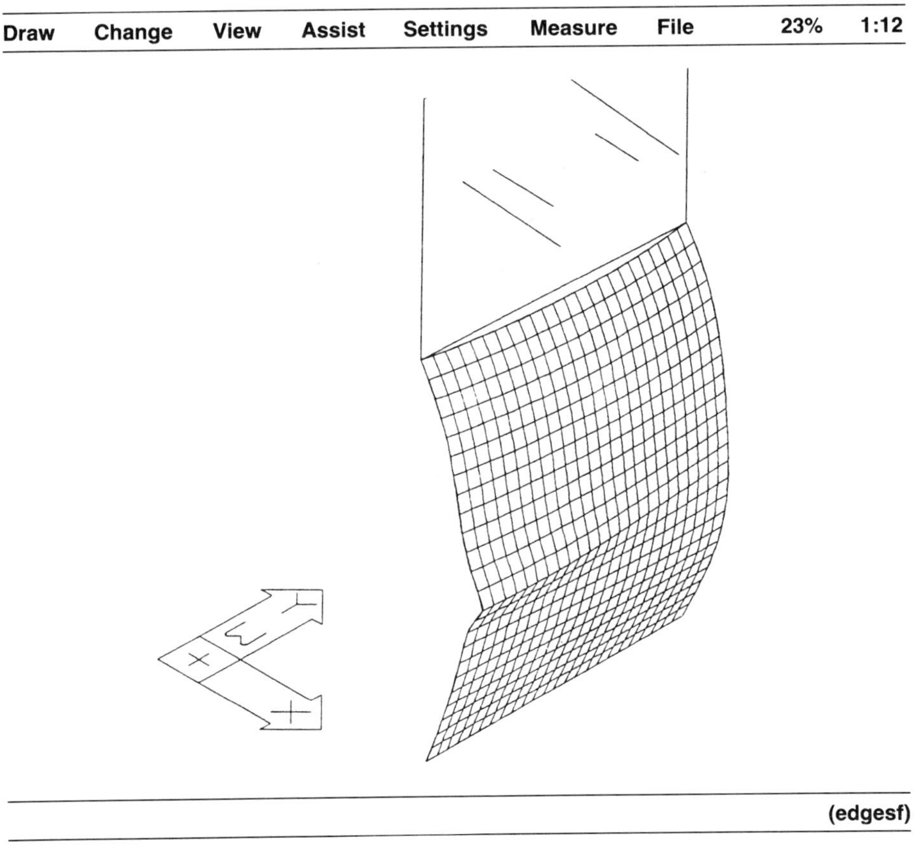

Figure 4.25 EDGESURF example.

appears at the top of the screen allowing you to dynamically select the second angle.

The D option allows perspective viewing. Selecting the perspective view entails two elements: the D option and the focal length of the lens of the camera. You can type in the distance or select it from a slider bar. You can adjust the focal length of the lens by selecting the Z option. This allows you to adjust your field of view by asking for a length specified in millimeters. A small lens length increases your view angle and creates the effect of looking at a solid with a wide angle lens.

To practice the other options of DVIEW, type **OFF** to get out of perspective and then you can try PAn (reposition the solid), TWist (rotate solid), Hide (drop wireform lines), Undo (delete last option), and eXit (leave DVIEW).

4.14 POINT SET STRUCTURES

Four examples from Table 4.2 are displayed as Figures 4.20 to 4.23.

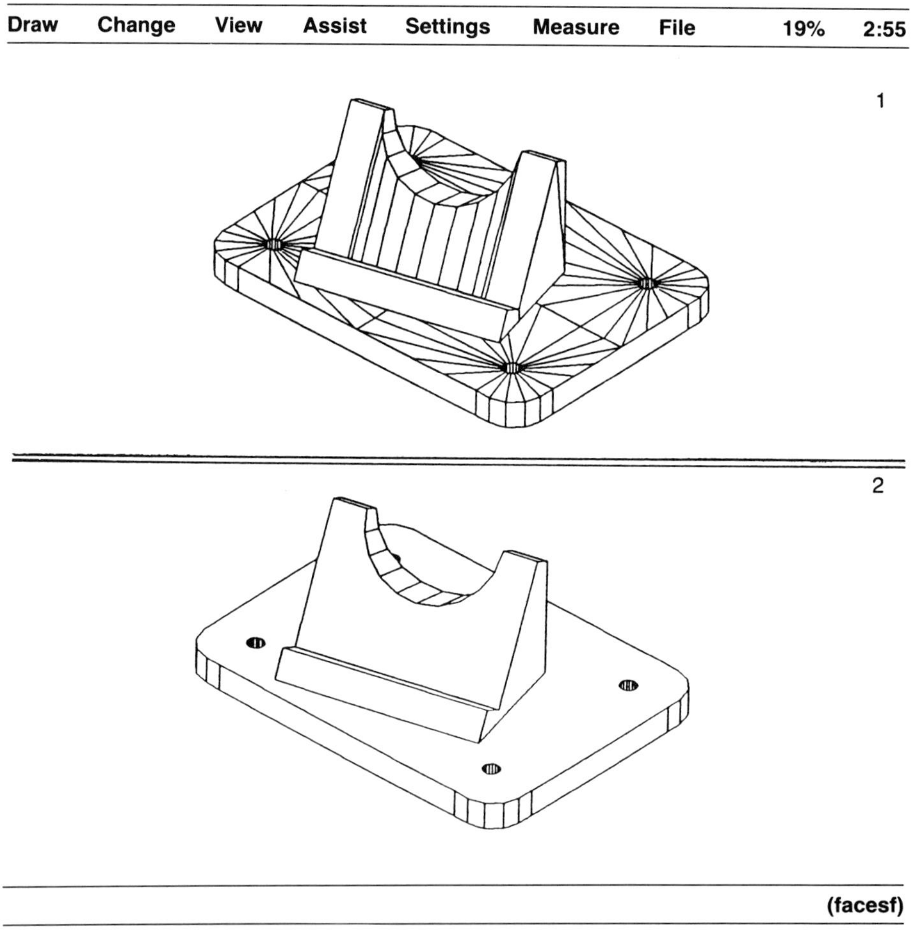

Figure 4.26 3DFACE example.

4.15 REPLICATIVE STRUCTURES

Replicative structures define entities of the same dimensionality as the original. As these do not depend on a parameterization, they can be presented in data tables as shown in Tables 4.3 and 4.4.

4.16 TOPOLOGICAL STRUCTURES

Topological structures provide a vehicle for representing various relationships that exist among Geometric, Replicative, and Special entities for solid constructions. The most useful of these are vertex (Fig. 4.13), TABSURF (Fig. 4.24), EDGESURF (Fig. 4.25), and 3DFACE (Fig. 4.26).

4.17 CHAPTER PROBLEMS

The problems appearing at the end of this chapter have been selected to introduce the basic types of broad, general problems encountered by an engineer or engineering technician using a CADD display system. These problems emphasize the interdisciplinary aspects of constructive geometry, which require a number of related pieces of CADD software. The software used for the production of problems in this chapter can be ordered from Autodesk, Inc.

PROBLEM 1 *Points in solids*

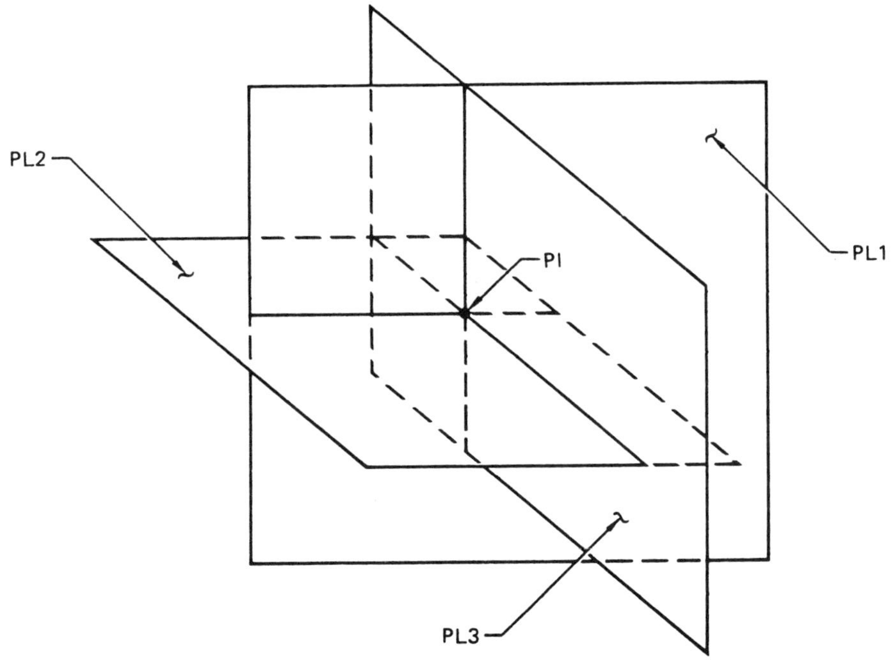

PROBLEM 2 *Circles in solids*

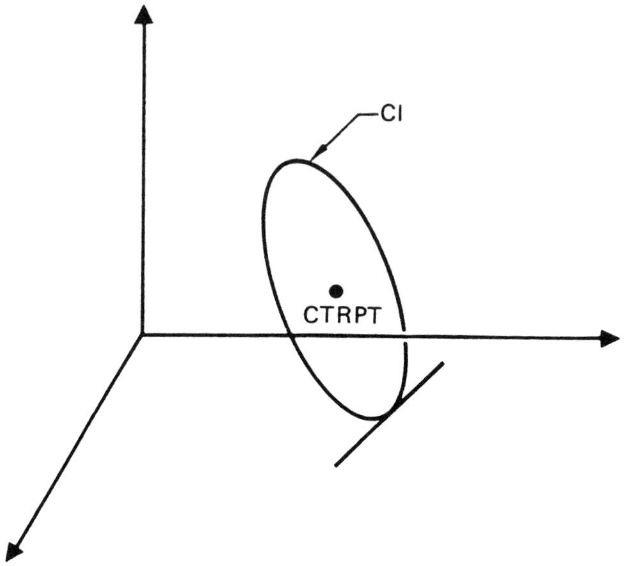

PROBLEM 3 *Lines in solids*

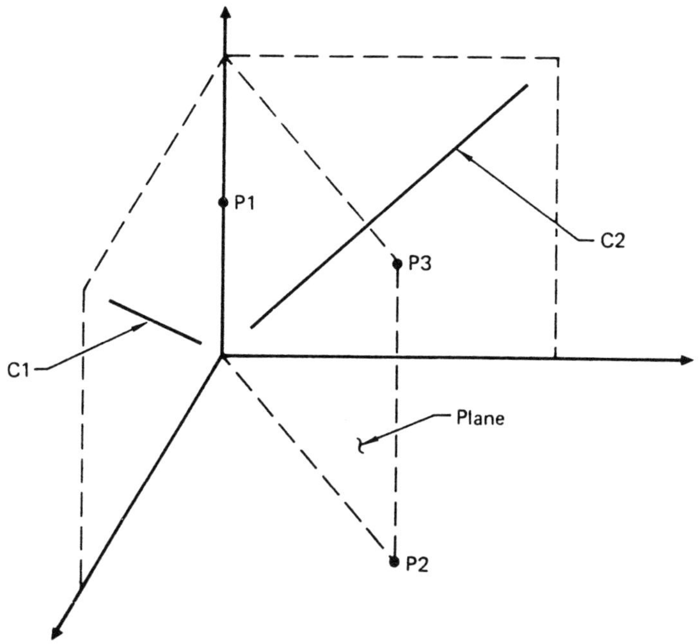

PROBLEM 4 *Arcs in solids*

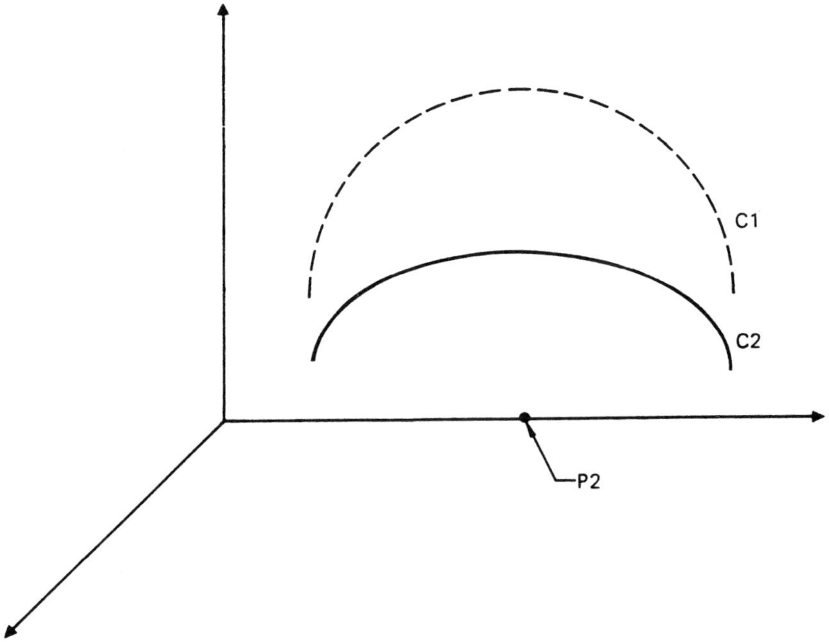

PROBLEM 5 *Curves in solids*

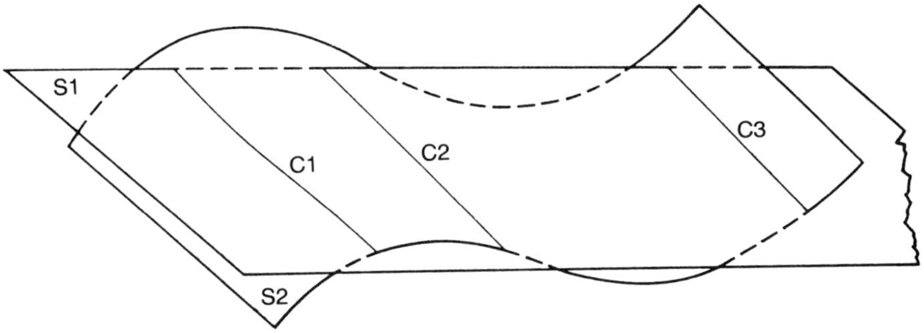

PROBLEM 6 *Splines in solids*

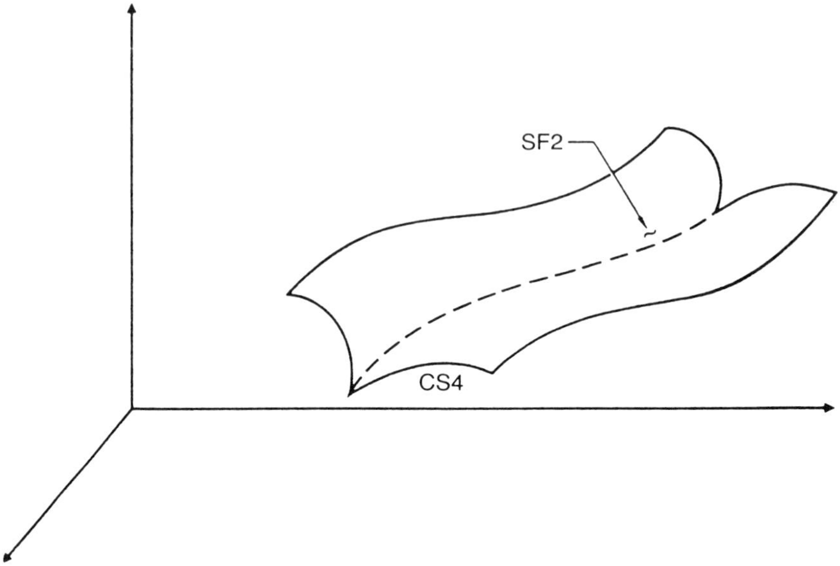

PROBLEM 7 *Translation of solids*

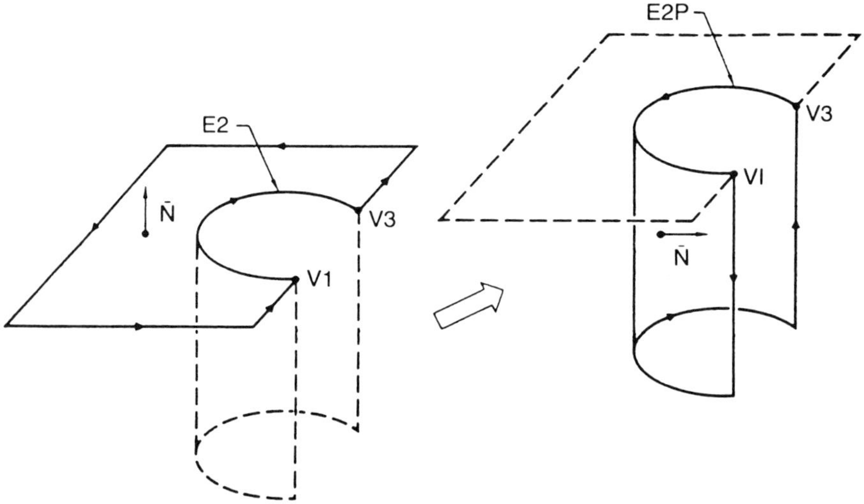

PROBLEM 8 *Rotation of solids*

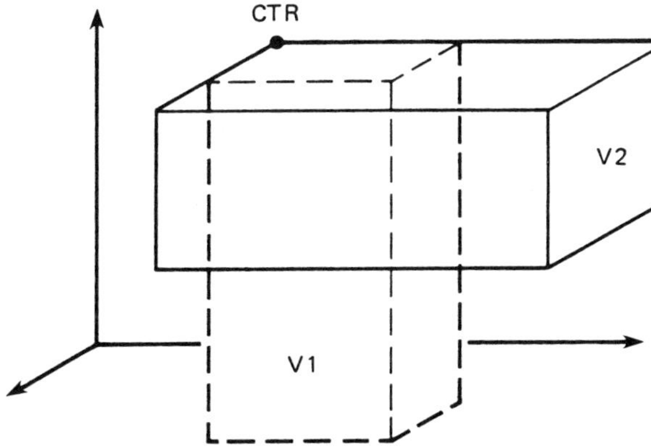

PROBLEM 9 *Reflection of solids*

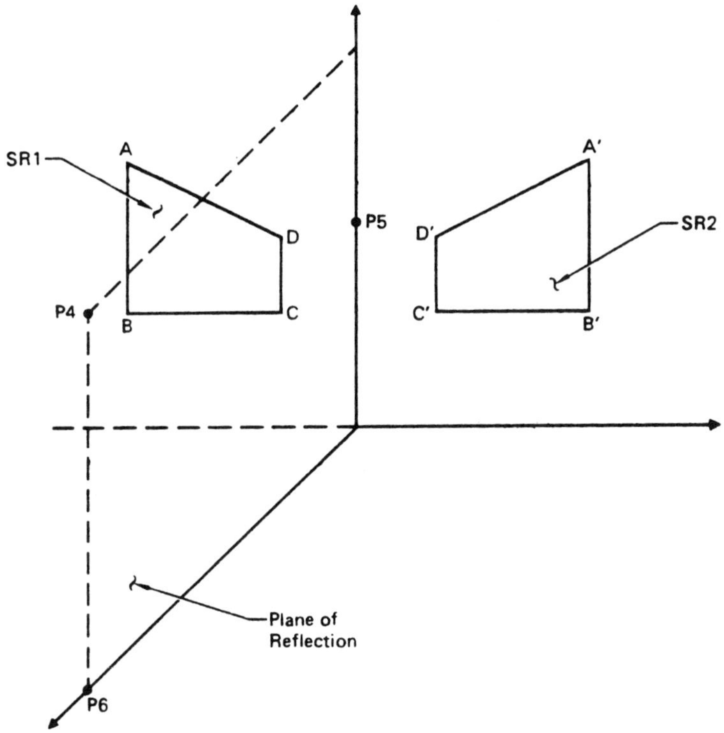

PROBLEM 10 *Scaling of solids*

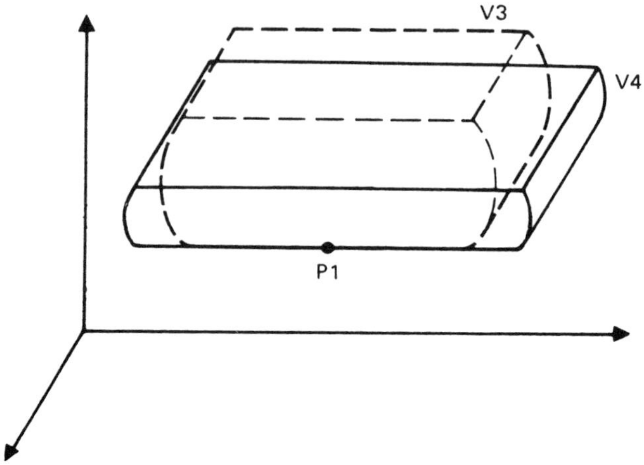

PROBLEM 11 *Dimension-selecting of solids*

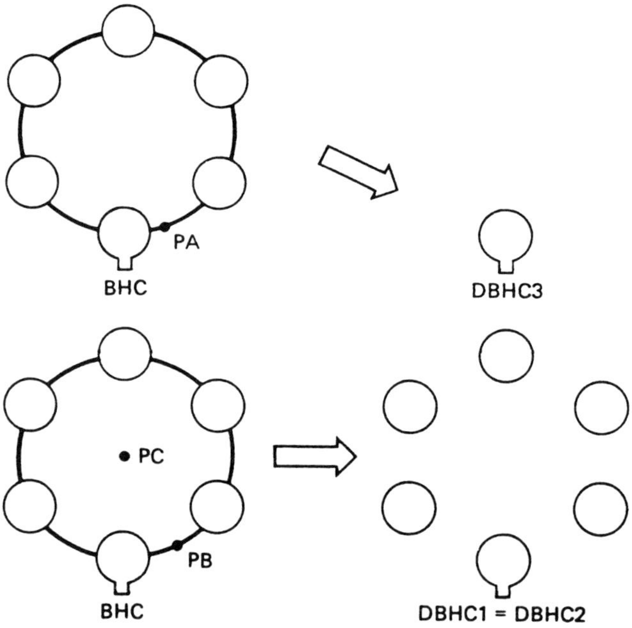

PROBLEM 12 *Union of solids*

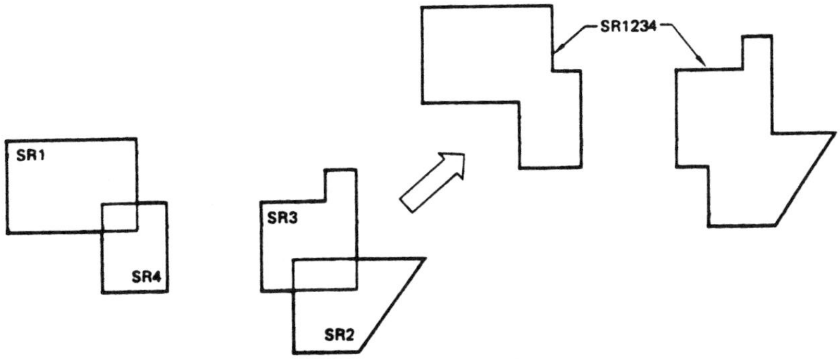

PROBLEM 13 *Intersection of solids*

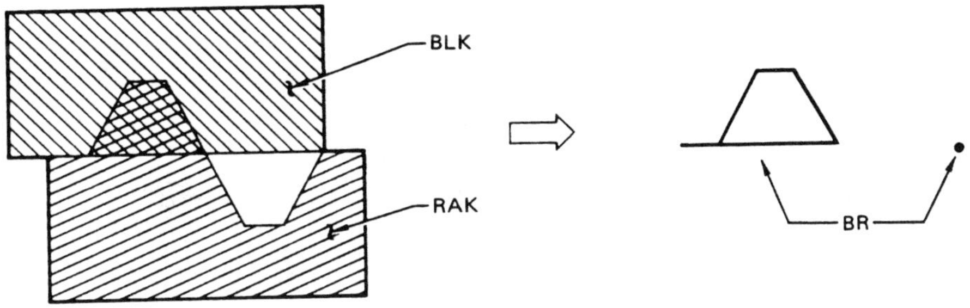

PROBLEM 14 *Difference of solids*

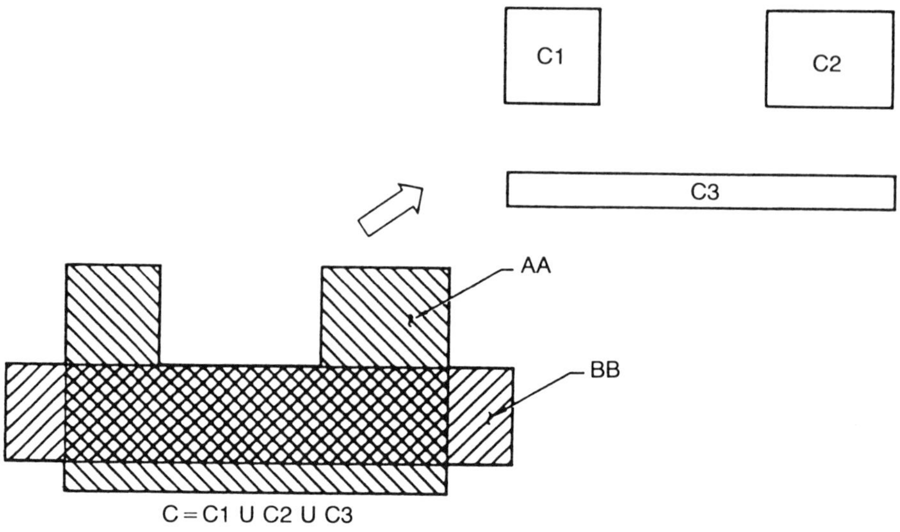

PROBLEM 15 *Replication of solids*

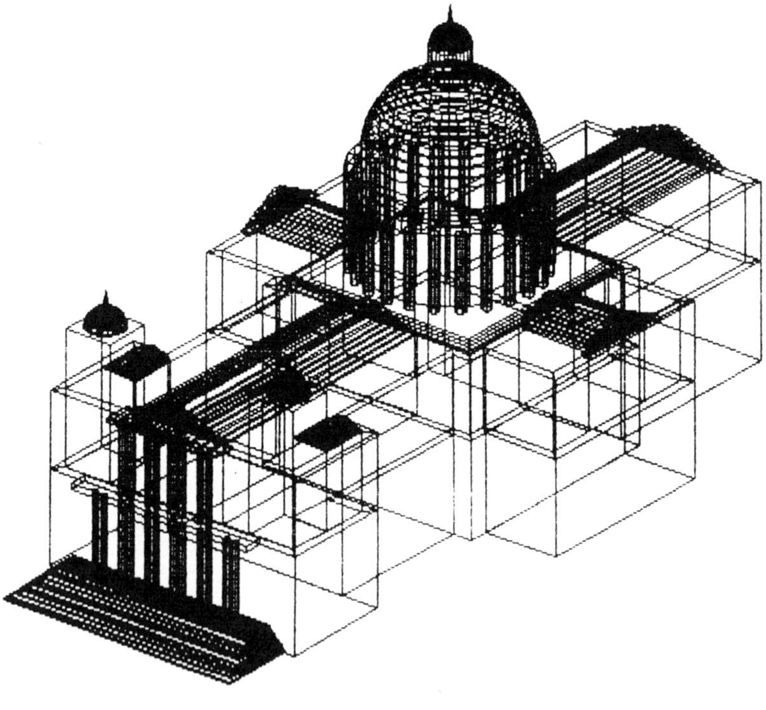

5 Simple Point/Line/Plane Relationships

This chapter contains several CAD problem situations that most often arise during the constructive geometry process. The series of problems contain rather simple point/line and plane relationships. Each of the problem situations is stated as a section heading; under each heading is the problem analysis, solution, and CADD construction required.

5.1 DETERMINE THE PROJECTIONS OF A LINE THROUGH A GIVEN POINT

Analysis

The projections of a line are made up of the projections of all points of the line. The projections of a point that lies on a line must, therefore, be found on the corresponding projections of the line.

Solution

Two projections displayed from or through the respective projections of the given point to any other given or assumed point describe the required line.

CADD Construction

Given: Point *A*, Figure 5.1.

Required: Any line through point *A*.

Point *B* *(X,Y,Z)* is displayed as shown in Figure 5.1. The corresponding projections of *A* and *B* are then connected. The direct projections (Fig. 5.1 case 1) of *AB* describe a line that passes through point *A*, since *A* is a point of the line. An alternative procedure displays a coordinate model containing points *A* and *B* in relationship to the base line *(BL)* as shown in case 2.

Point *b* coordinates are assumed to be in the fourth quadrant in case 2. Begin the display by locating b' (step 1), projecting downward $(-Y)$ from *BL* (step 2) until you locate *b* (step 3). Next connect b' and a' (step 4), then *a* and *b* (step 5), and last verify *a* to a' (step 6).

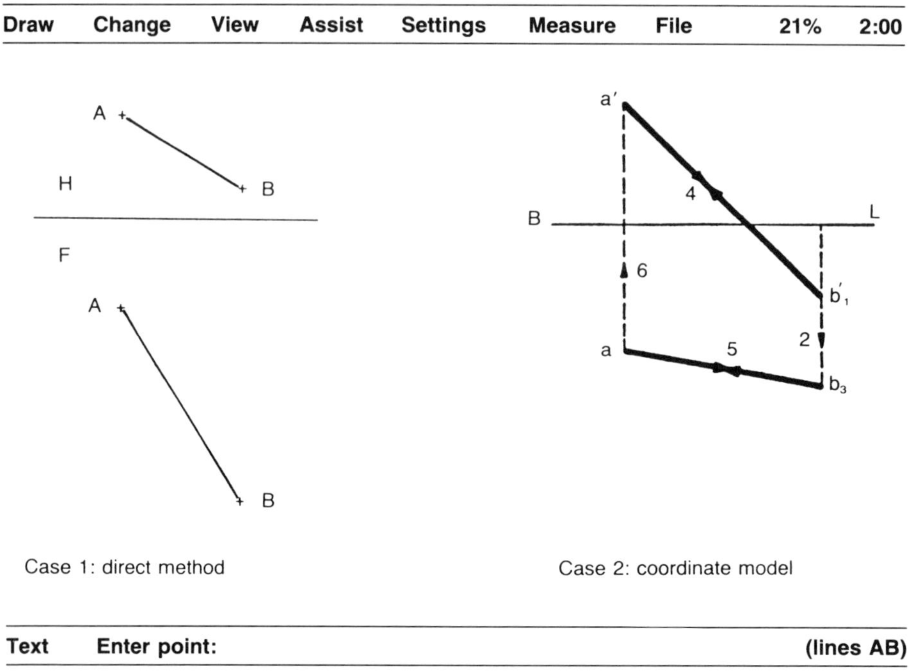

Figure 5.1 CADD problem 1.

5.2 DETERMINE LINE PROJECTION

Analysis

A line is defined by any two points on it. The lines joining the respective projections of two points constitute the projections of the line through those two points. All the qualities of a line can be determined from the display: length, location, and direction.

Solution

The projections of two points, *A* and *B* in Figure 5.2, can be displayed such that the lines joining them describe the required line.

CADD Construction

Required: To determine a line parallel to *H* and oblique to a vertical intersecting plane of projection, extending from point *A* located in the first quadrant above *H* to point *B* located in the second quadrant. Case 1 of Figure 5.2 shows the direct method of projection while case 2 shows the *V* projection of a point in the first quadrant above *H*. Point *A* is then displayed above *BL* as a', and a is located vertically below it and at any distance below *BL* (steps 1 to 3). The *V* projection of the line is horizontal, and both projections of point *b* are above *BL*, since *B* is in the second quadrant (steps 4 to 6).

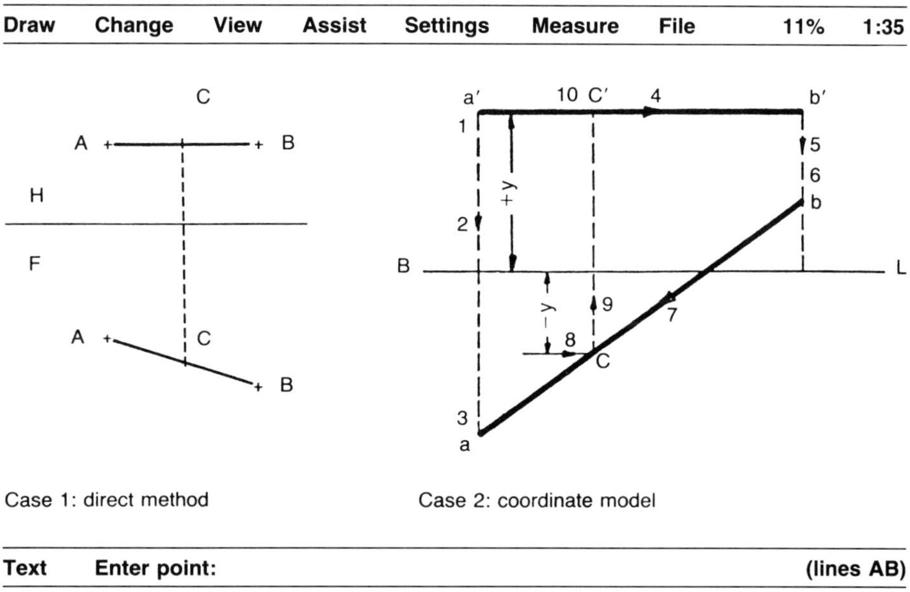

Figure 5.2 CADD problem 2.

5.3 DETERMINE POINT PROJECTIONS ON A LINE

Analysis

If a point lies on a given line in space, its projections lie on the corresponding projections of the line and are vertically above and below each other as shown by point *C* in Figure 5.2.

Solution

The location of a third point in Figure 5.2 is indicated by the positions of the lines of sight indicated in case 1 and the intersection of the *BL* (steps 9 and 10) in case 2.

Construction

Given: Line *AB*, Figure 5.2.

Required: A third point *C* on *AB*. The *H* projection of point *C* lies on the *H* projection of *AB* and is below the *BL* (step 8). The *V* projection is on *a′b′* and vertically above (steps 9 and 10).

5.4 FIND THE *HVP* PIERCING POINTS OF A COORDINATE MODEL

Analysis

This problem assumes no direct method solution because of the infinite direct viewing space. In a coordinate model, however, it can be assumed that *HVP* are definitely located by *BL* and *PL*.

A new user of coordinate modeling is advised to designate the *HVP* piercing points of all lines as points *H, V,* and *P*. The *V* projections of all points in *H* lie

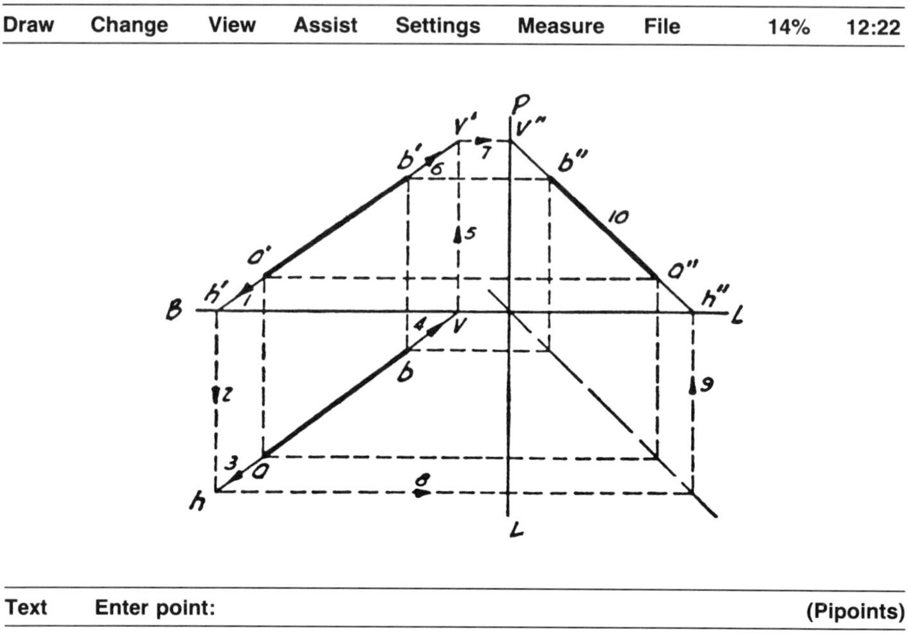

Figure 5.3 CADD problem 3.

on *BL*, and the *V* projections of all points of a given line lie on the *V* projection of the line. The *V* projection of the point common to a given line and *H* is therefore the intersection of the *V* projection of the line with *BL*.

Solution

The intersection of the *H* and *V* projections of a given line with the *BL* is the *H* projection of the *V* piercing point and the *V* projection of the *H* piercing point, respectively. The intersections of vertical lines through these points with the other projections of the line are the other projections of the desired piercing points.

CADD Construction

Given: Line *AB* in Figure 5.3.

Required: *H* and *V* piercing points of *AB*.

The intersection of $a'b'$ extended and *BL* gives h'—the *V* projection of the *H* piercing point (step 1). A perpendicular to *BL* at h' intersects *ab* at *h;* the *H* projection of the *H* piercing point is step 2 and 3. The extension of *ab* intersects *BL* at *v;* the *H* projection of the *V* piercing point is step 4. The intersection of a perpendicular to *BL* at *v* with $a'b'$ locates v', and the *V* projection of the *V* piercing point is steps 5 and 6.

5.5 DISPLAY A PLANE

A plane whose position and direction are not specified can be displayed as shown in Figure 5.4 by producing any two intersecting (case 1) or parallel lines (case 2).

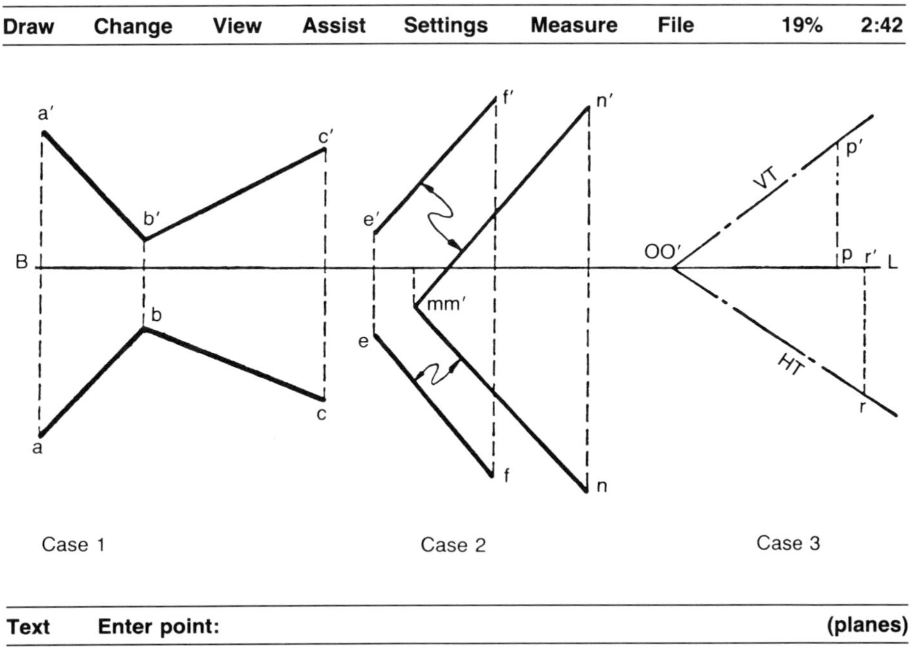

Figure 5.4 Display plane.

If desired or convenient, the two intersecting lines may be the intersections of the plane with the coordinate planes (case 3), that is, traces, which must either intersect on *BL* or be parallel to it.

5.6 DISPLAY LINE IN A PLANE

Analysis

A plane is defined by any two lines in it (Chapter 3). A line lies in a plane if it has two points in common with it. A line through or connecting any two points on the two given lines must lie in the plane that they define.

Solution

A line intersecting one of the given lines and parallel to the other given line will lie in the plane that they define and can therefore be displayed.

CADD Construction

Given: Planes *T* shown in Figure 5.5.

Required: Lines in the respective planes.

If the plane is displayed by traces (case 1 or 2), the points *v* and *h* are displayed as steps 1 and 2 and steps 3 and 4. These are chosen anywhere on the traces and are respective piercing points of the displayed line in the coordinate planes. If a plane is defined by any two lines (case 3), any two points, say *a* and *c* or *d* and *e*, may be connected by display steps 1 to 4 and steps 5 and 6.

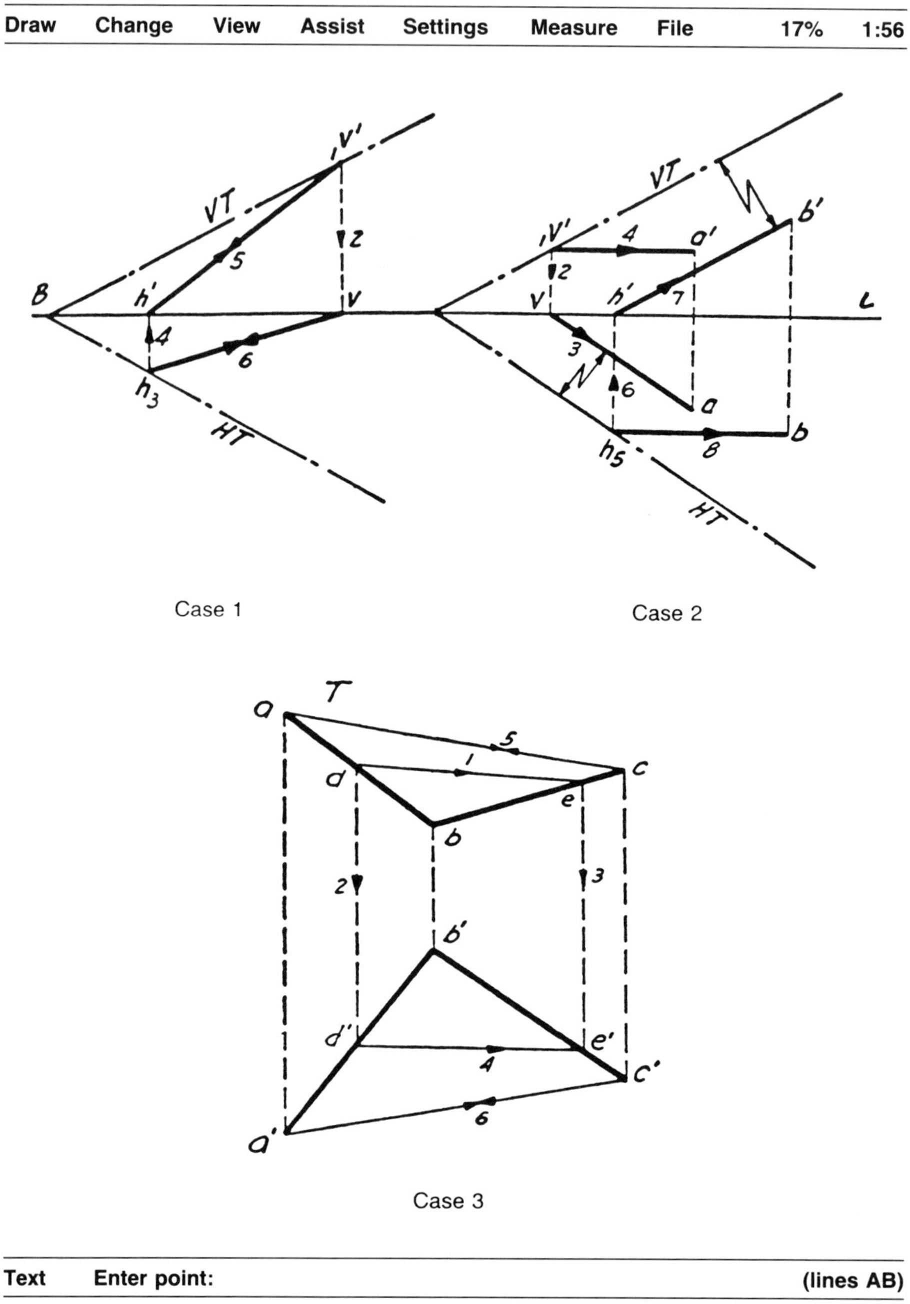

Figure 5.5 CADD problem 5.

If a plane is displayed by any three points as shown in Figure 5.6 (case 1), then a polygon of any number of sides has one projection and may be displayed following the steps shown in case 1. Any horizontal line in a given plane is called a *horizontal* of that plane as shown in case 2 of Figure 5.6. Here *VA* of plane *T* is that horizontal component and can be displayed as in case 2. Case 3 demonstrates the fact that traces have directions. The trace direction can be determined by displaying a horizontal component *ae* and *cd* by steps 1 through 6.

5.7 DISPLAY POINT IN A PLANE

Analysis

A plane is displayed by any two lines in it. A point is known to lie in a plane if it lies on a line of the plane. A line lies in a plane if it has two points in common with it.

Solution

Any point on either of the two lines that define a plane, or upon any line joining points of these two lines, lies in the given plane as displayed in Figure 5.6.

5.8 DETERMINE THE TRACES OF A PLANE DISPLAYED BY INTERSECTING LINES

Analysis

The traces of a plane are the loci of the corresponding piercing points of all lines of the plane.

Solution

If the vertical, horizontal, and profile piercing points of the given lines are displayed, they will define the corresponding traces of the plane. The *V* and *H* traces will intersect on the *BL* (Figure 5.7, case 1) or be parallel to it (case 2). The *V* and *P* traces will intersect on the *PL* or be parallel to it.

CADD Construction

Given: Lines *ab* and *cd*, Figure 5.7.

Required: Traces of the plane of *AB* and *CD*.

By displaying Figure 5.7, the projections of the *V* piercing points of *AB* and *CD* are *v,v'* (steps 1 to 3) and *v,v'* (steps 4 to 6). The projections of the *H* piercing points are *h,h'* (steps 8 to 10) and *h,h'* (steps 11 to 13). The *V* projection of the *V* trace of the required plane *T* passes through *V'* at step 7; its *H* projection lies on *BL* through *V*.

5.9 DISPLAY THE TRACES OF A PLANE BY POINT AND LINE

Analysis

The plane of a given line and a given point as in Figure 5.8 is displayed by the given line and a line connecting the given point with any point of the line. The

Draw Change View Assist Settings Measure File 22% 11:23

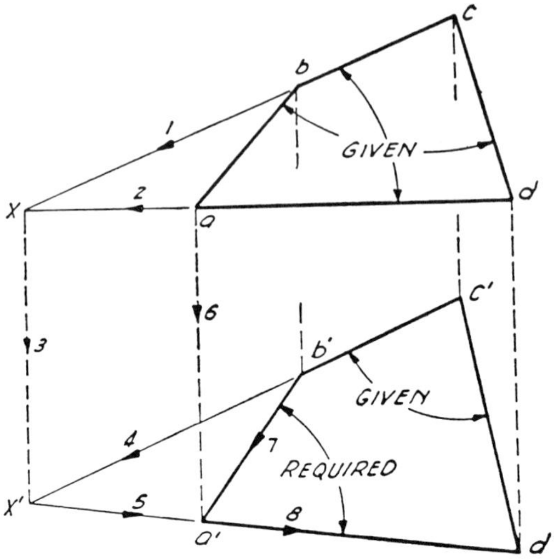

Case 1

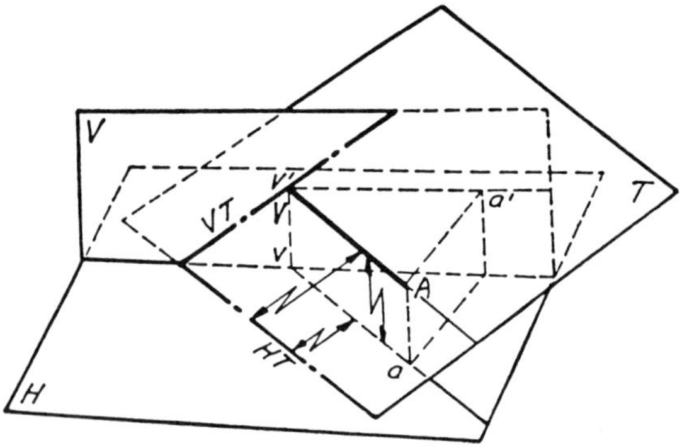

Case 2

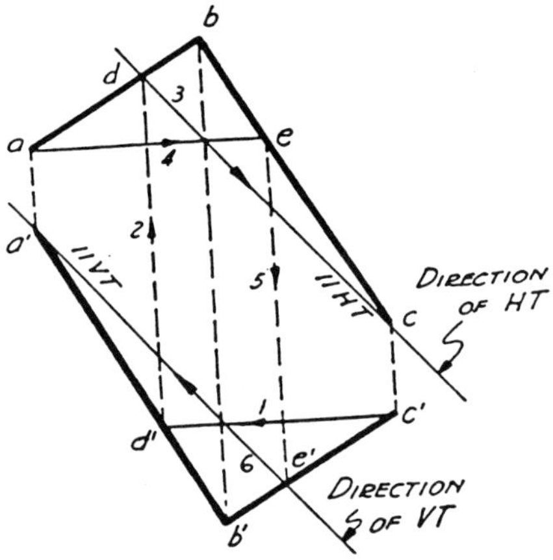

Case 3

Text Enter point: (planeT)

Figure 5.6 CADD problem 5 continued.

Draw **Change** **View** **Assist** **Settings** **Measure** **File** **24%** **1:45**

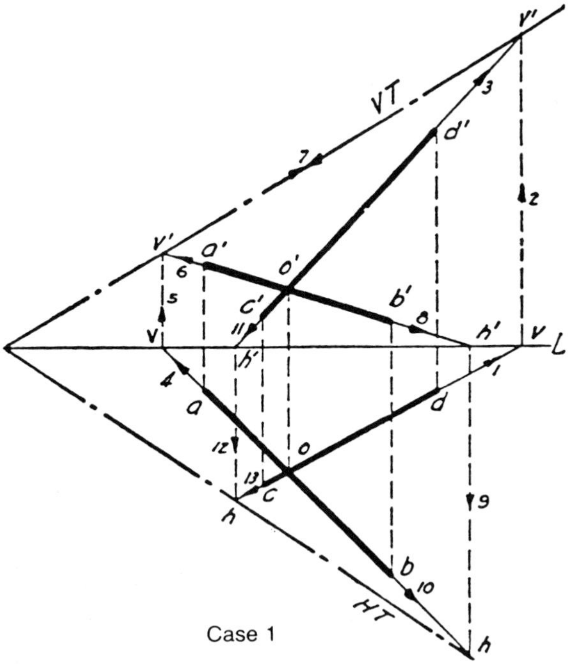

Case 1

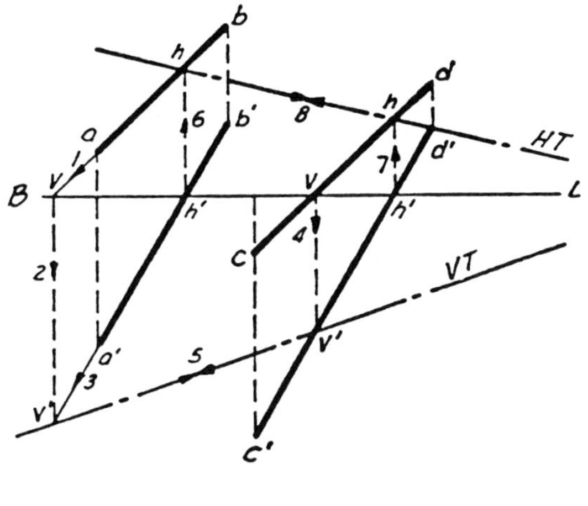

Case 2

Text **Enter point:** **(planeT)**

Figure 5.7 CADD problem 8.

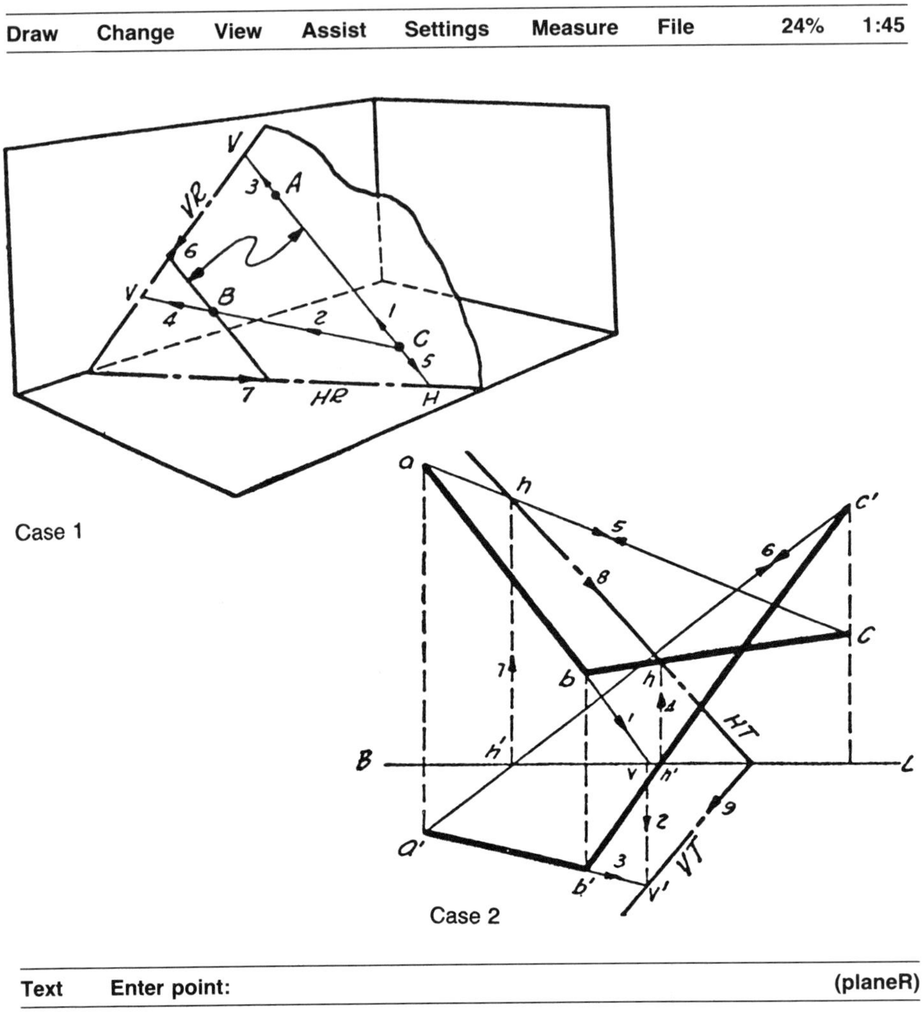

Figure 5.8 CADD problem 9.

plane can be displayed also by the given line and a line through the point parallel to the given line.

Solution

The traces of the plane of the given point and line displayed in Figure 5.8 as line *AB* and point *C* will pass through the respective piercing points of the given line and the auxiliary parallel or intersecting line.

5.10 DISPLAY THE TRACES OF A PLANE BY THREE POINTS

Analysis

Three points can be used to display a plane because they define two or three intersecting lines joining them as shown in Figure 5.9.

Draw Change View Assist Settings Measure File 24% 1:45

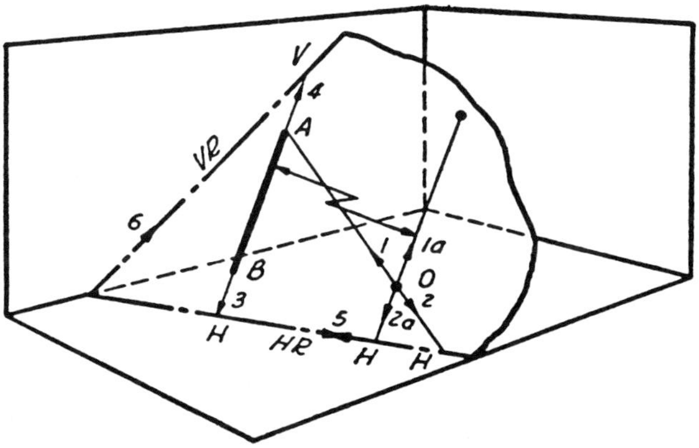

Text Enter point: (planeR)

Figure 5.9 CADD problem 10.

Solution

The required traces pass through the respective piercing points of the lines joining the given points as displayed in Figure 5.9.

CADD Construction

Given: Line *AB* and point *O*.

Required: Display construction line parallel to *AB* (steps 1 and 2) and connect *AO, BH, AV*. Display *HR* (step 5) and *VR* (step 6).

5.11 DISPLAY LINES PARALLEL TO PLANES

Analysis

The only requirement that a line needs to fill in order to be parallel to a plane is that it be parallel to any line of that plane.

Solution

Three cases can be displayed to describe this solution (Fig. 5.10). In case 1 a line is constructed parallel, case 2 shows a line parallel to *BL,* and case 3 shows parallel inside traces.

CADD Construction

Given: Plane *ABC* (step 1), *OPS* (step 2) and *R* (step 3).

Required: Display construction line parallel to *ABC (de), BL (r′),* and *m,m′* (step 3). Display steps shown in Figure 5.10.

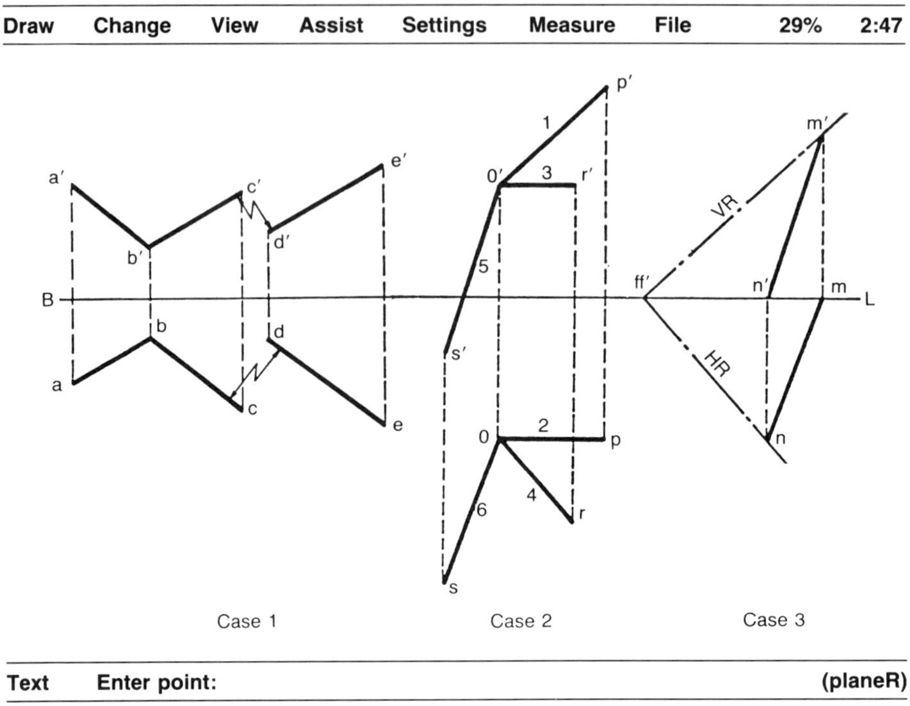

Figure 5.10 CADD problem 11.

5.12 DISPLAY PLANES PARALLEL TO PLANES

Analysis

If two intersecting lines of plane *R* are parallel to two intersecting lines of plane *S*, they are parallel as shown in Figure 5.11.

Solution

Three cases can be displayed to describe this solution.

CADD Construction

Given: Plane *ABC/R* (step 1), *EFG/S* (step 2) and *R/S* (step 3).

Required: Display plane *R* parallel to *ABC*, *S* parallel to *EFG*, and trace *R* parallel to trace *S*. Display steps are shown in Figure 5.11.

5.13 DISPLAY LINE PARALLEL TO PLANE AND LINE

Analysis

The two lines are displayed non-parallel and non-intersecting. A plane is to be displayed parallel to the given line.

Solution

A geometric model shown in Figure 5.12 (case 1) demonstrates this.

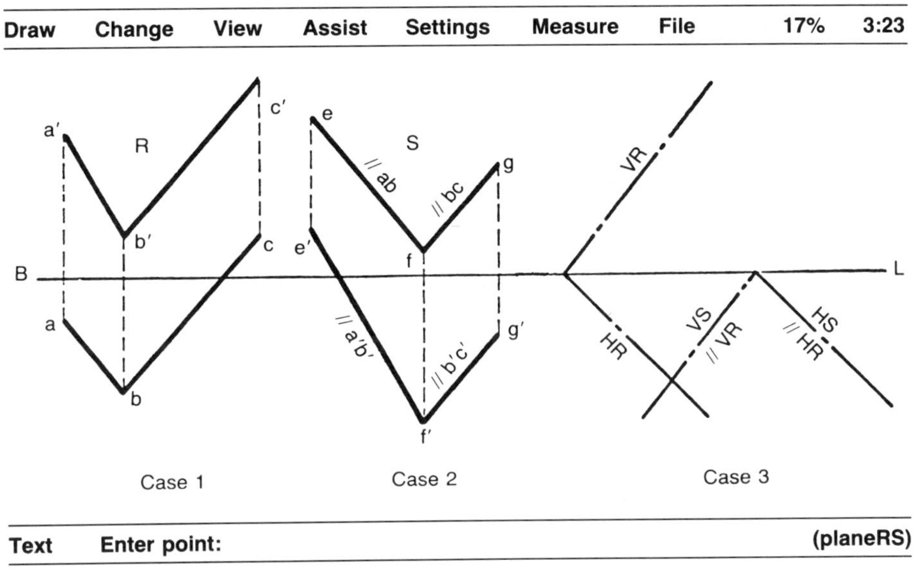

Figure 5.11 CADD problem 12.

CADD Construction

Given: Line *AB*.

Required: Display line *DC* parallel to *AB* plane surface. This relationship is displayed in case 2 through the 8-step procedure.

5.14 DISPLAY LINE PARALLEL TO PLANE THROUGH A POINT

Analysis

A plane passes through a given point if the point lies on a line of that plane. A plane is parallel to a given line if it contains a line parallel to the given line.

Solution

A geometric model shown in Figure 5.13 (case 1) demonstrates this.

CADD Construction

Given: Line *AB* and point *O*.

Required: Display line *AB* parallel to plane *R* through point *O*. Figure 5.13 is a display of the 8-step procedure.

5.15 DISPLAY PLANE PARALLEL TO TWO LINES

Analysis

A plane is parallel to two given lines if it contains a line parallel to the two given lines.

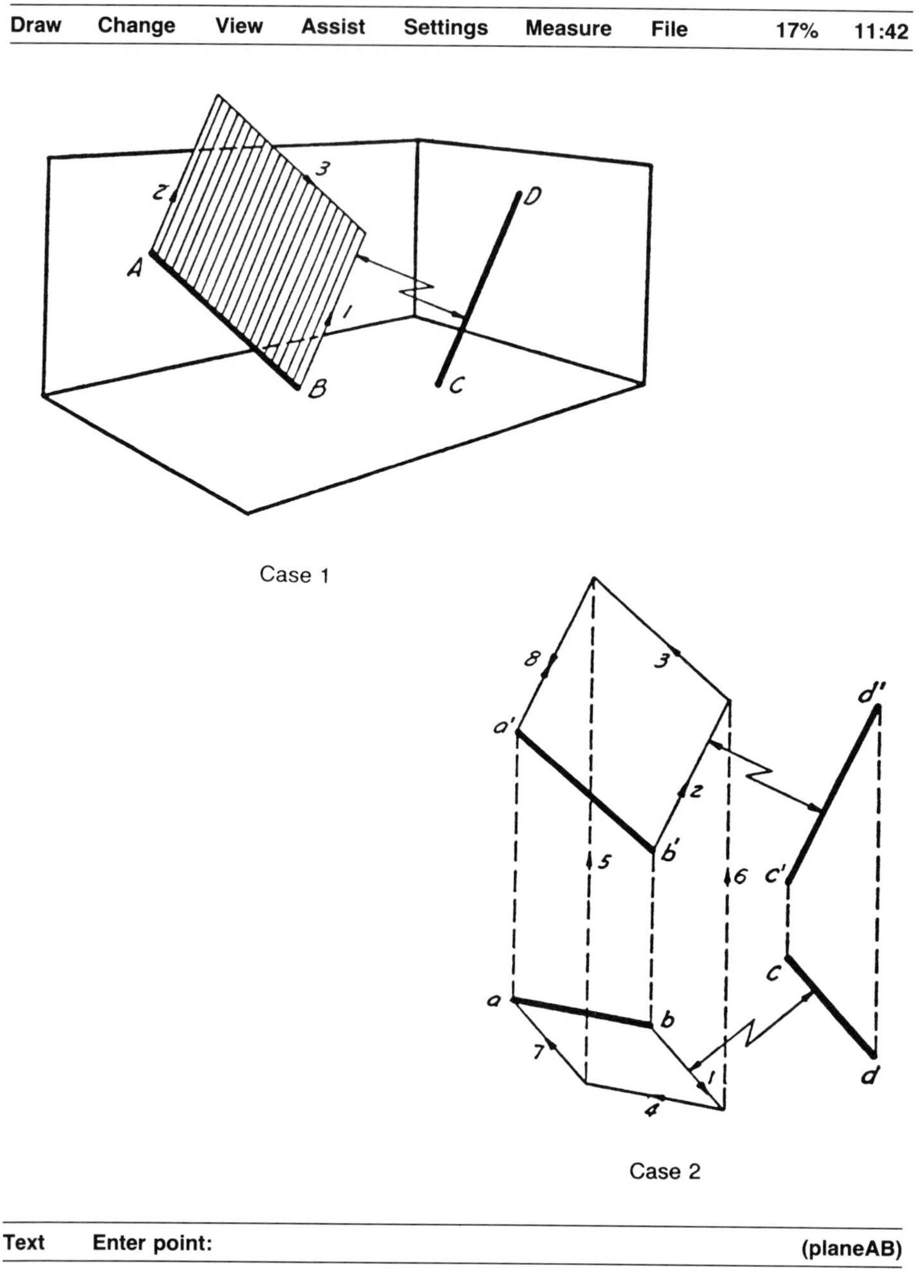

Figure 5.12 CADD problem 13.

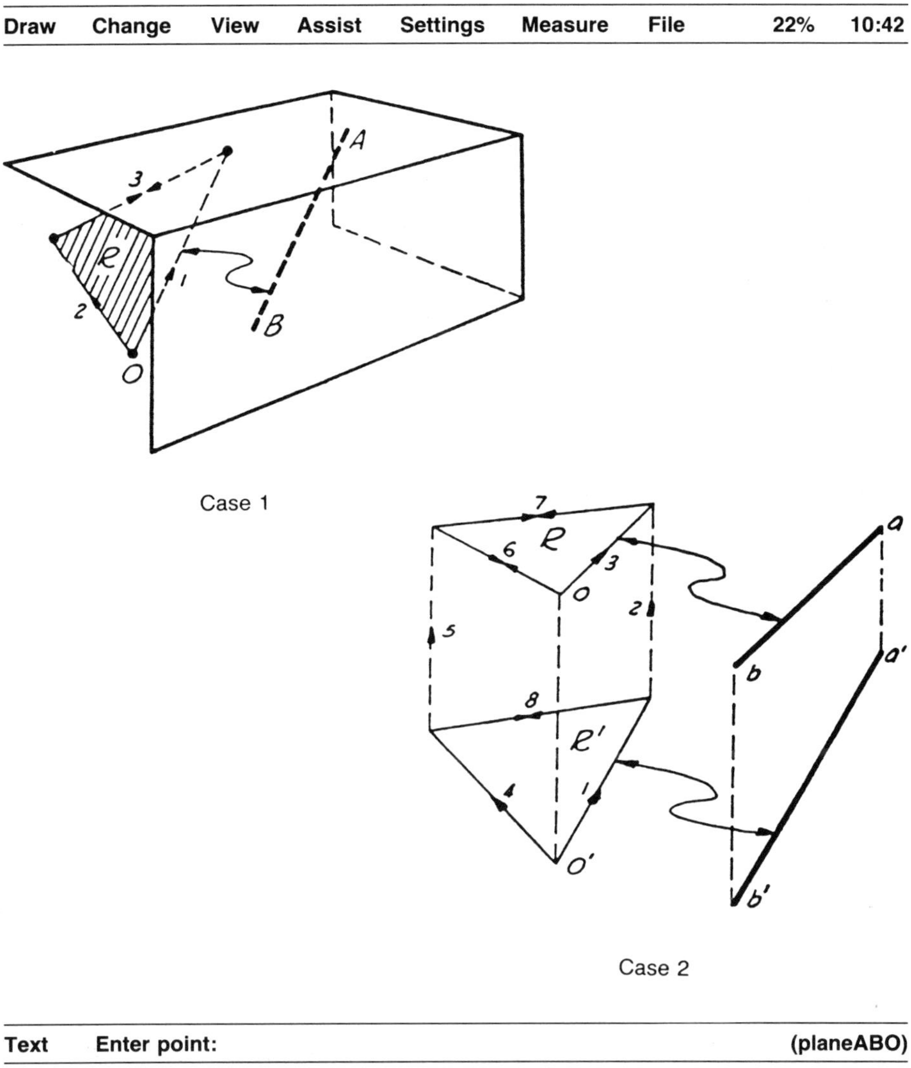

Figure 5.13 CADD problem 14.

Solution

A geometric model shown in Figure 5.14 (case 1) demonstrates this.

CADD Construction

Given: Plane *EOF*, lines *AB* and *CD*.

Required: Display plane *EOF* parallel to lines *AB* and *CD* as indicated in Figure 15.14 (case 2), which shows the 7-step procedure.

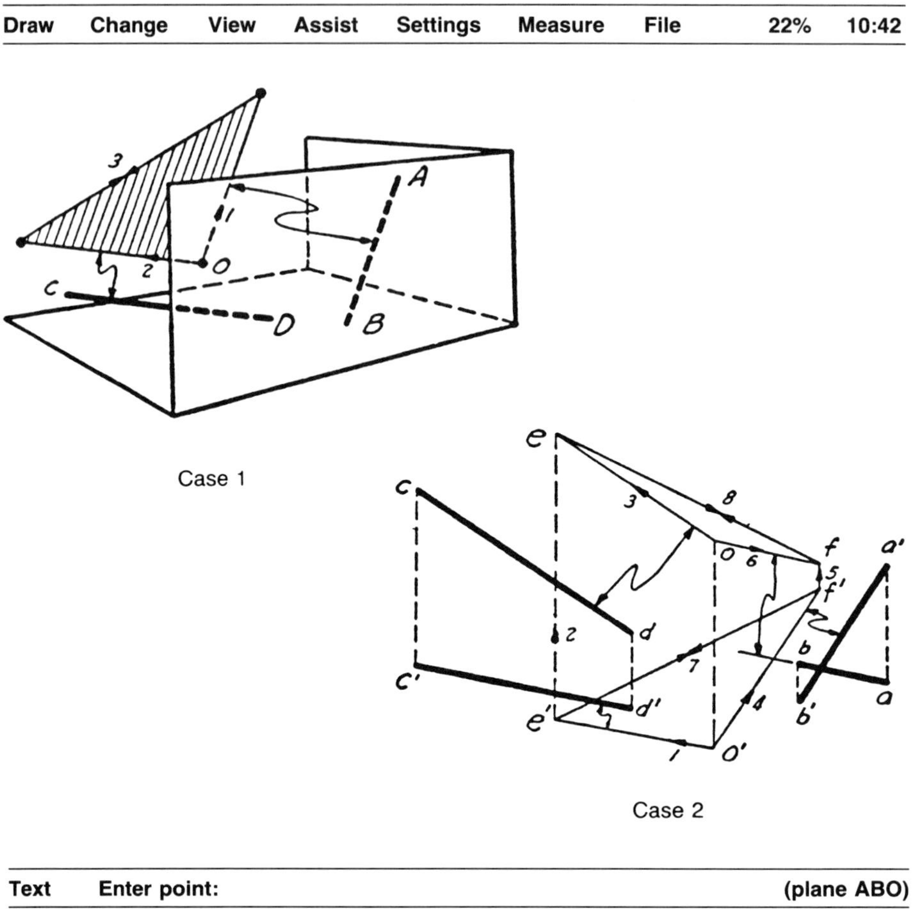

Figure 5.14 CADD problem 15.

5.16 DISPLAY PLANE PARALLEL TO PLANE THROUGH A POINT Analysis

A plane passes parallel through a point if the point lies on a line of the plane parallel to a second plane.

Solution

A geometric model shown in Figure 5.15 (case 1) demonstrates this.

CADD Construction

Given: Planes *MNO, ABC,* and point *O*.

Required: Display plane *ABC* parallel to plane *MNO* through point *O*. Figure 15.15 (case 2) shows the 6-step procedure.

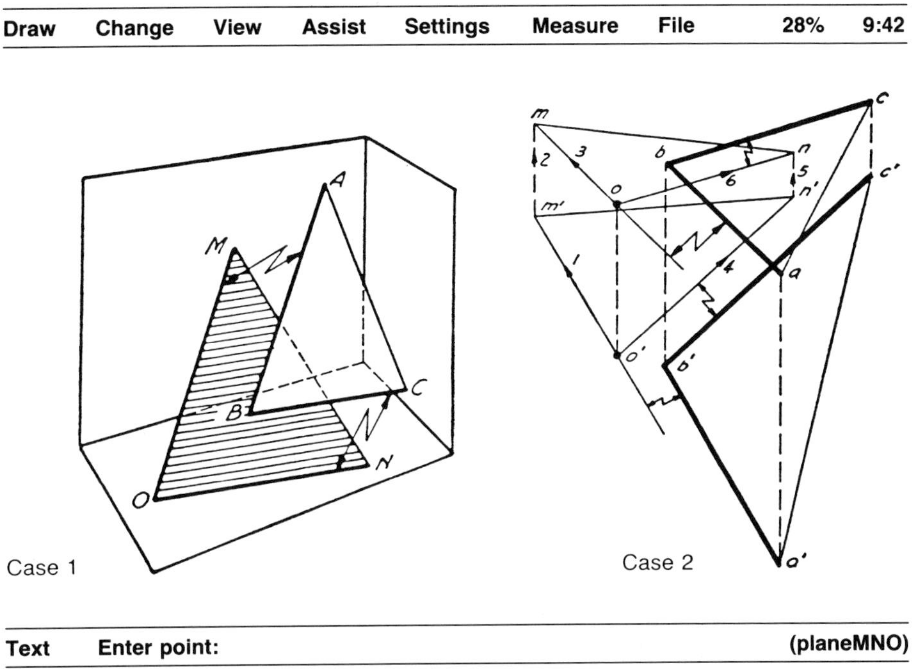

Figure 5.15 CADD problem 16.

5.17 DISPLAY LINES PERPENDICULAR TO PLANES

Analysis

A line is perpendicular to a plane if it is perpendicular to two lines in that plane.

Solution

A geometric model shown in Figure 5.16 (case 1) demonstrates this.

CADD Construction

Given: Planes *MNO*, *ABC*, and point *O*.

Required: Display plane *ABC* parallel to plane *MNO* through point *O*. Figure 15.15 (case 2) shows the 6-step procedure.

5.18 DISPLAY PLANE PERPENDICULAR TO PLANE

Analysis

Two planes are perpendicular if one of them has been displayed to contain a line that is perpendicular to the other plane.

Solution

A geometric model shown in Figure 5.17 (case 1) demonstrates this.

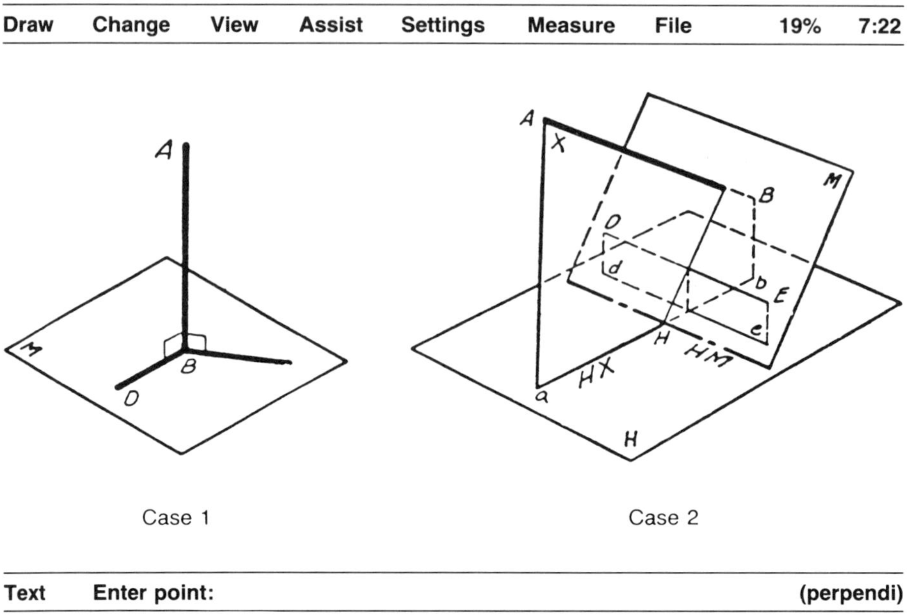

Figure 5.16 CADD problem 17.

CADD Construction

Given: Planes *S* and *T*.

Required: Display plane *T* perpendicular to *S* by making *DE* in plane *T* perpendicular to *S* as shown in the 11-step procedure (Fig. 5.17, case 2).

5.19 DISPLAY PLANE PERPENDICULAR TO LINE

Analysis

To pass a plane through a given point perpendicular to a given line requires a plane that is perpendicular to a given line intersecting that line at a point. The given line is perpendicular to each line in the plane passing through the point of intersection. The condition of perpendicularity can be seen between the given line and only three of the lines of the plane. These are the horizontal line, a line parallel to the vertical, and a line parallel to the profile. A plane through a given point and parallel to any plane that is perpendicular to a given line is also perpendicular to that line.

Solution

If through any point of the given line, as in Figure 5.18 (case 1), the lines are displayed perpendicular to the line, they constitute a plane perpendicular to the line. A plane through the given point parallel to this plane is the required plane. The traces of this plane (case 2) are perpendicular to the corresponding projections of the given line.

Draw **Change** **View** **Assist** **Settings** **Measure** **File** **19%** **7:22**

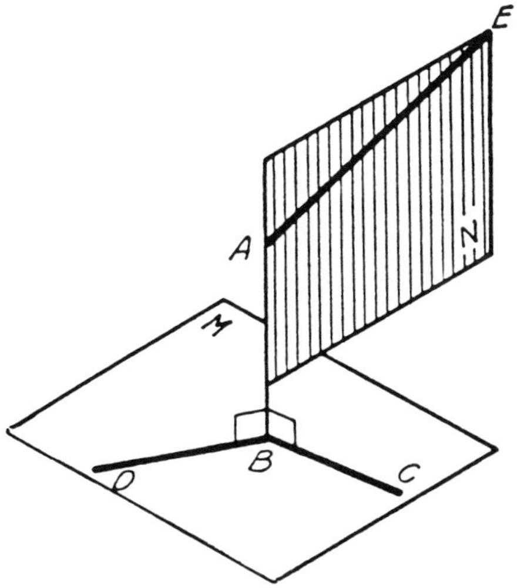

Case 1

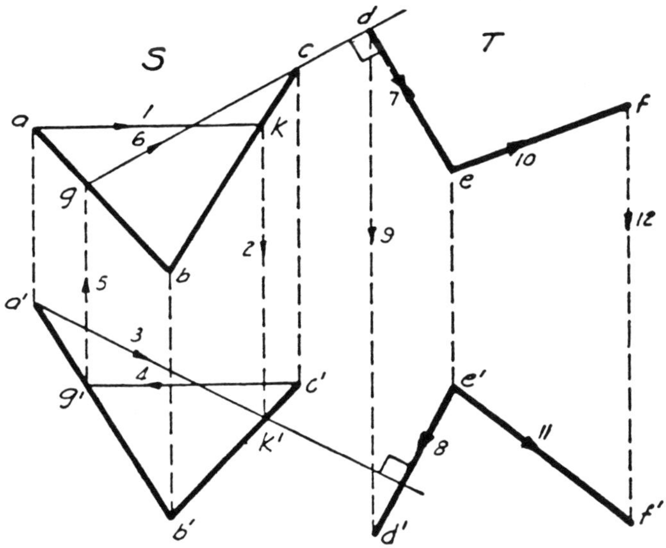

Case 2

Text **Enter point:** **(perpendi)**

Figure 5.17 CADD problem 18.

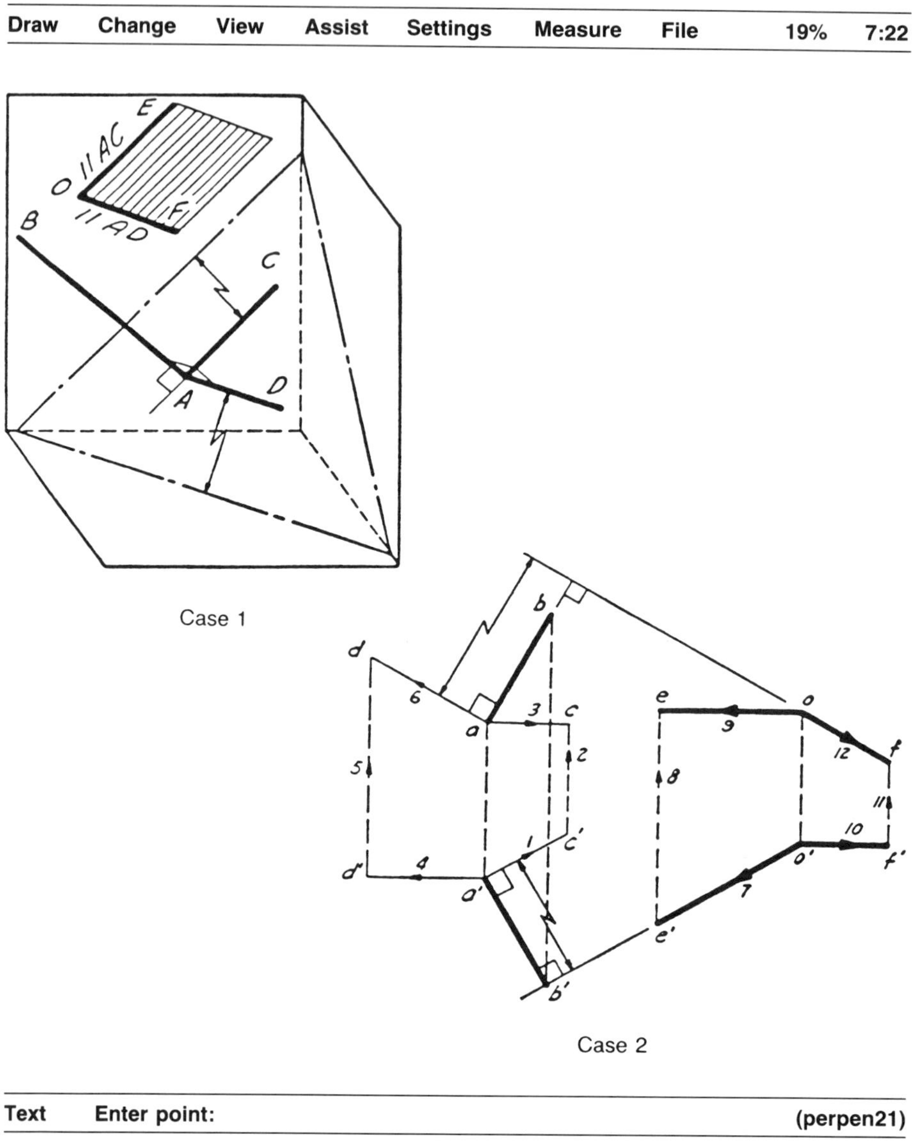

Figure 5.18 CADD problem 19.

CADD Construction

Given: Line *AB* and point *O* (Fig. 5.18).

Required: A plane through point *O* perpendicular to line *AB*. Through point *A* two lines are displayed perpendicular to *AB* (case 2). Line *AC* (steps 1 to 3) is displayed parallel to the vertical, and therefore the angle of 90 degrees shows on the vertical plane. Line *AD* (steps 4 to 6) is displayed horizontal, and the angle of 90 degrees shows on the horizontal plane. Plane *DAC* is perpendicular to *AB*. Lines

OE and *OF* are displayed through *O* parallel, respectively, to *AC* (steps 7 to 9) and *AD* (steps 10 to 12). Plane *EOF* is parallel to plane *DAC* and hence is perpendicular to *AB*. It is apparent that lines *OE* and *OF* could have been displayed through *O* without passing lines *AD* and *AC* through *A* at all because *OE* is a vertical of the required plane and *e'o'* (step 7) is therefore perpendicular to *a'b'* and *OF* is a horizontal of the required plane. That makes *fo* (step 12) perpendicular to *ab*.

5.20 DETERMINE DISPLAY PLANE, POINT, AND PLANE

Analysis

A display plane can be constructed through a point if the point lies on a line of a second plane perpendicular to the given plane. A plane *S* is perpendicular to a second plane *T* if *S* contains a line that is perpendicular to *T* (Fig. 5.19, case 1). A line is perpendicular to a plane if it is perpendicular to two intersecting lines of that plane. A line is known to be perpendicular to a plane if its respective projections are perpendicular to the directions of the corresponding traces of the plane shown in case 2.

Solution

If a line *OF* is displayed perpendicular to the given plane through the given point *O*, any plane passing through line *OF* as shown in Figure 5.19 (case 1) will be perpendicular to the given plane.

CADD Construction

Given: Point *O* and plane *T (ABC)* (Fig. 5.19, case 2).

Required: A plane through point *O* perpendicular to plane *T*. *CD* (steps 1 to 3) is a horizontal plane *ABC*. *AE* (steps 4 to 6) is a vertical of *ABC*. The *H* projection *dc* (step 3) of *DC* and vertical *a'e'* (step 6) of *AE* are the directions of the *H* and *V* traces of *ABC*. The projections of a line through *O* perpendicular to *ABC* are displayed *fo* and *f'o'* (steps 7 and 8). Any other line *OM* (steps 9 and 10) through *O* defines with *OF* a plane *FOM* through *O* perpendicular to *ABC*.

5.21 DISPLAY INTERSECTION OF LINES

Analysis

Two nonparallel lines will intersect each other if extended. This extension is often outside the display model. When this happens, it is sometimes useful to display the point in which a given line pierces the horizontal or vertical model boundary. In the case of two nonparallel lines, two piercing points are displayed.

Solution

The intersection of the *H* projections of the given lines shown in Figure 5.20, case 1, is the *H* projection of the point at which either line pierces the *H* projecting plane of the other.

CADD Construction

Given: Lines *AB* and *CD* (Fig. 5.20, case 1).

Required: The point *P* at which *AB* pierces *H* projection of *CD*. The *H* trace

Draw **Change** **View** **Assist** **Settings** **Measure** **File** **19%** **7:22**

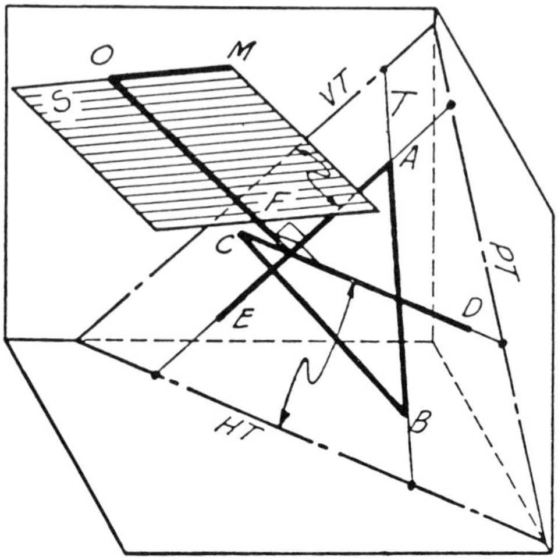

Case 1

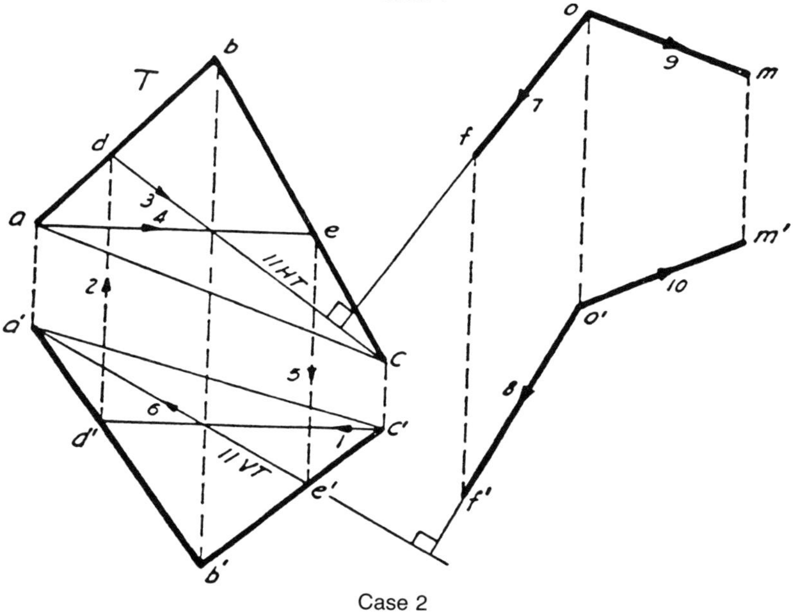

Case 2

Text **Enter point:** **(perpen21)**

Figure 5.19 CADD problem 20.

Draw Change View Assist Settings Measure File 19% 7:22

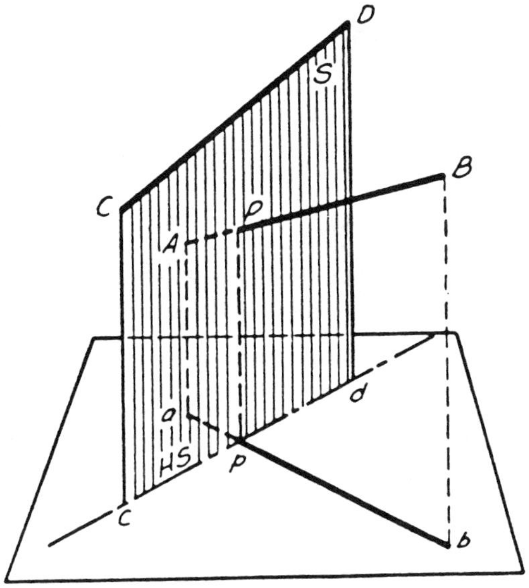

Case 1

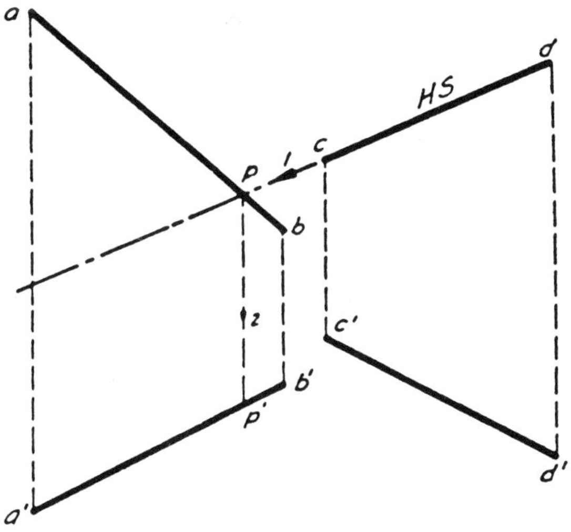

Case 2

Text Enter point: (piercing)

Figure 5.20 CADD problem 21.

of the *H* projection plane of *CD* coincides with *cd* (step 1) shown in case 2. The *H* projection of point *P* at which *AB* pierces *S (H)* is at *p* and *V* is at *p'*.

5.22 DISPLAY INTERSECTION OF LINE AND PLANE

Analysis

The required point is common to the given line and the plane. The common point is always displayed as the intersection of two lines. The second line in this case must lie in the given plane.

Solution

The required piercing point is the point of intersection of the given line and the line of intersection of the given plane with the orthogonal *(HVP)* plane of the given line as shown in case 2 of Figure 5.21.

CADD Construction

Given: Line *AB* and plane *T* (Fig. 5.21, case 1).

Required: Piercing point of line *AB* in plane *T*. The *H* projecting plane *S* is displayed through *AB* (case 2). Its *H* trace coincides with *ab*. *CD* and *DE* pierce *S* at points *M* and *N* (steps 1 and 2). The projections of the line of intersection *MN* of *S* and *CDE* are therefore *m'n'* and *mn* (step 3). The orthographic projection of *AB* intersects *m'n'* at *p'* (the piercing point). The fourth step of the display shows point *P* is at *p* on *ab* and *mn*.

5.23 DISPLAY INTERSECTION OF PLANE AND PLANE

Analysis

The intersection of two planes is a straight line displayed by two points that are common to both planes. If both planes are displayed as edge views (lines), the line of intersection will appear as a point.

Solution

If two planes are displayed by their traces, the *H* traces will intersect in one of the required points and the orthogonal view will display the intersection of the second point. The connection of these two points is the line of intersection of the two planes.

CADD Construction

Given: Planes *A* and *B* (Fig. 5.22, case 1).

Required: The line of intersection. Two orthogonal views are displayed in Figure 5.22 (case 2) that represent the vertical cutting plane *S* shown in case 1. This cutting plane is passed through plane *A* and plane *B* at *M, O, R* and *N;* this results in *P*(1). The traces of *S* and *T* (second cutting plane) are then used to display the line of intersection as shown in case 2. The 18-step display involves the orthographic views of plane *A* (points *ABCD*) and plane *B* (points *EFGH*) and the common line of intersection.

Draw **Change** **View** **Assist** **Settings** **Measure** **File** **19%** **7:22**

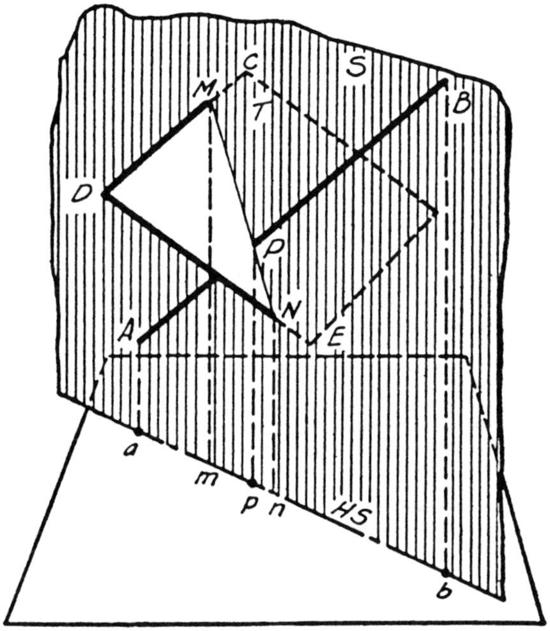

Case 1

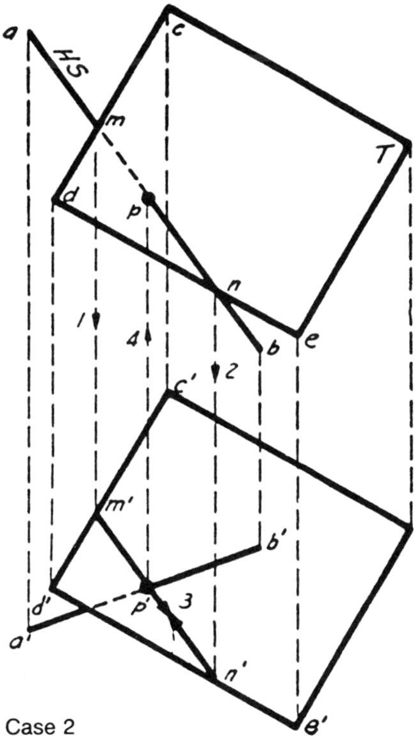

Case 2

Text **Enter point:** **(piercing)**

Figure 5.21 CADD problem 22.

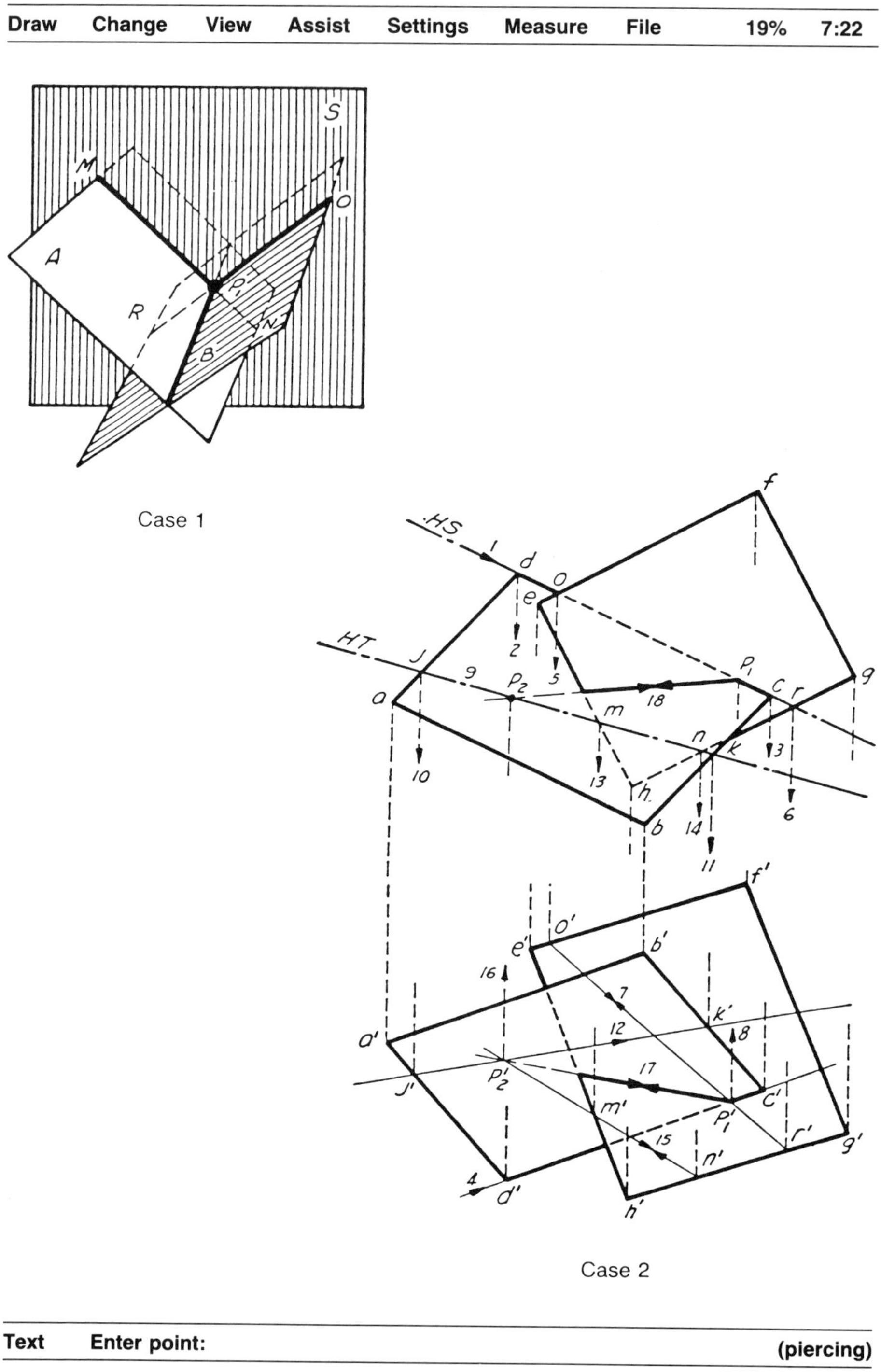

Figure 5.22 CADD problem 23.

5.24 CHAPTER PROBLEMS

The problems appearing at the end of this chapter have been selected to introduce the basic types of broad, general problems encountered by an engineer or engineering technician using a CADD display system. These problems emphasize the interdisciplinary aspects of constructive geometry, which require a number of related pieces of CADD software. The software used for the production of problems in this chapter can be ordered from Autodesk, Inc.

PROBLEM 1 *Point projection*

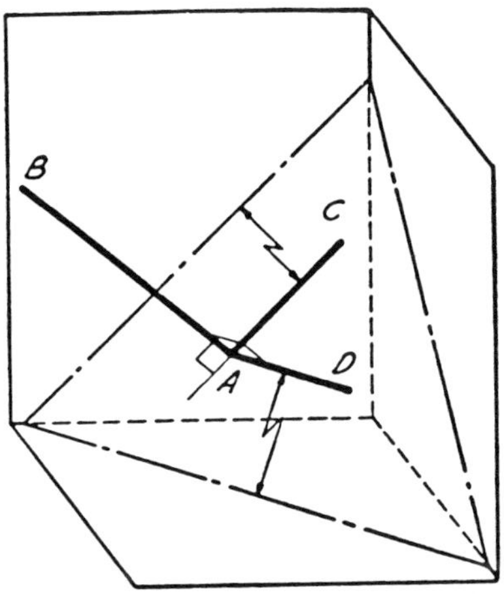

PROBLEM 2 *Line projection*

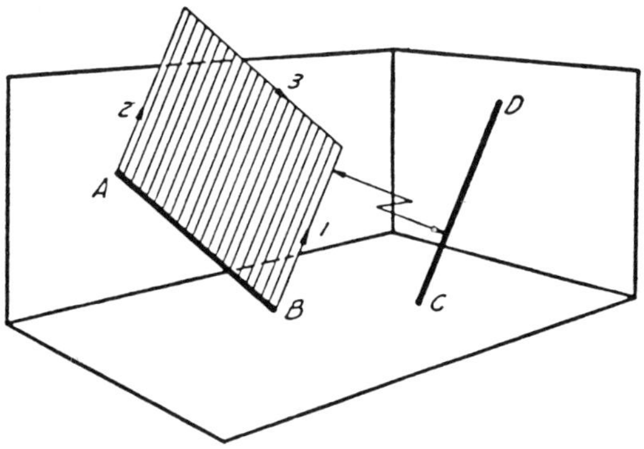

PROBLEM 3 *Point on a line*

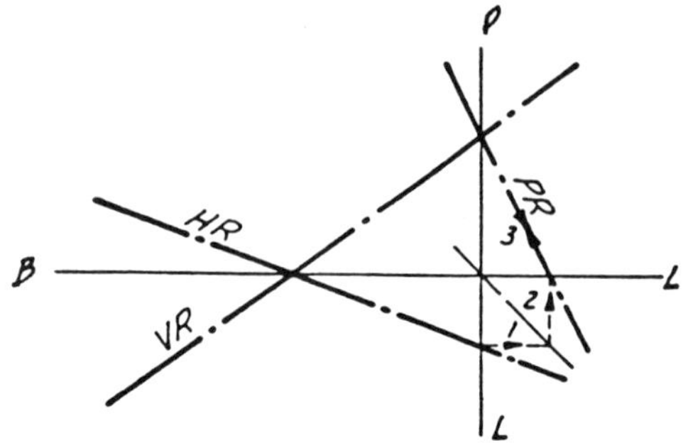

PROBLEM 4 HVP *projection*

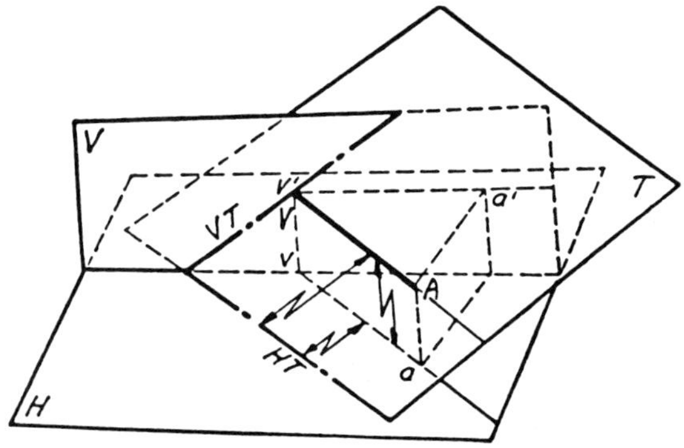

PROBLEM 5 *Plane projection*

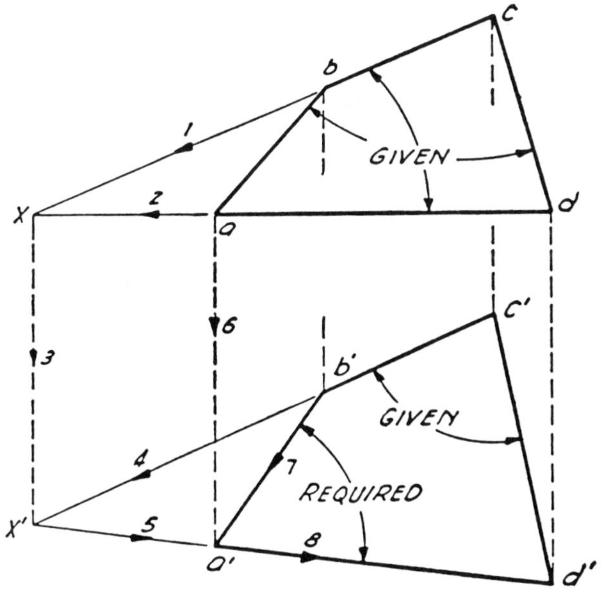

PROBLEM 6 *Line in a plane*

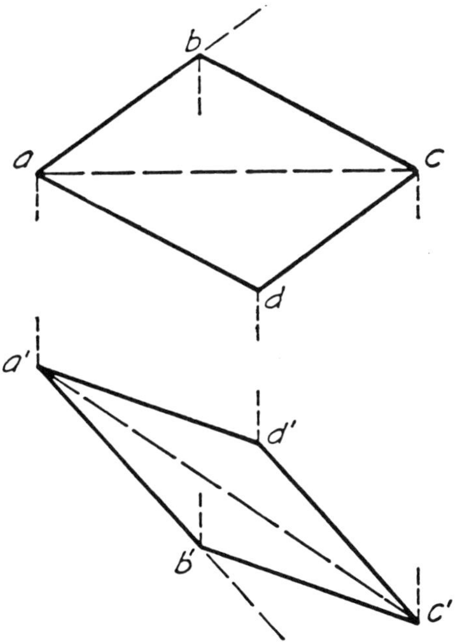

PROBLEM 7 *Point in a plane*

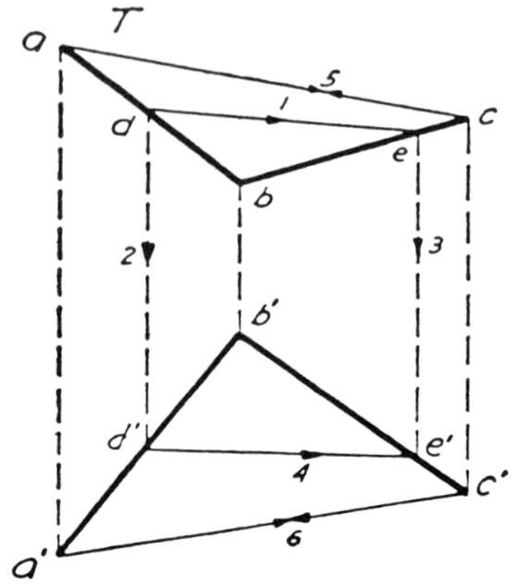

PROBLEM 8 *Trace of a plane*

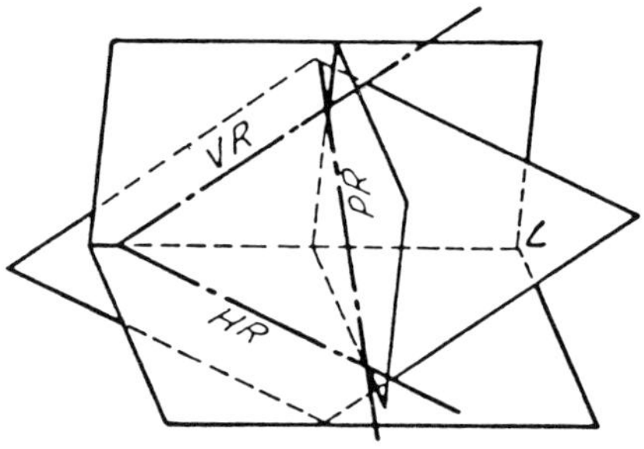

PROBLEM 9 *Line in a plane*

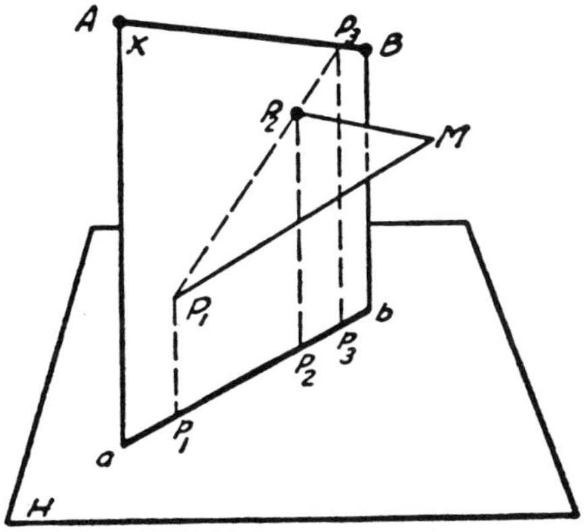

PROBLEM 10 *Line parallel to a plane*

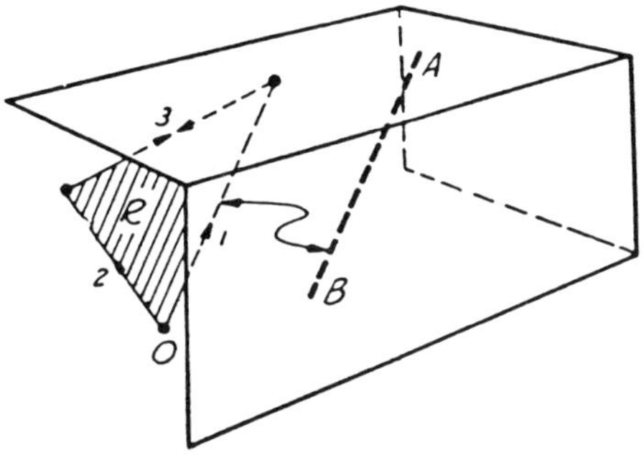

PROBLEM 11 *Plane parallel to plane*

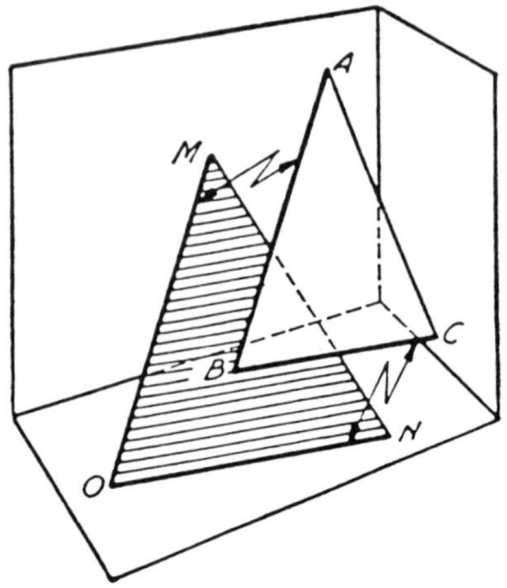

PROBLEM 12 *Line parallel to plane and line*

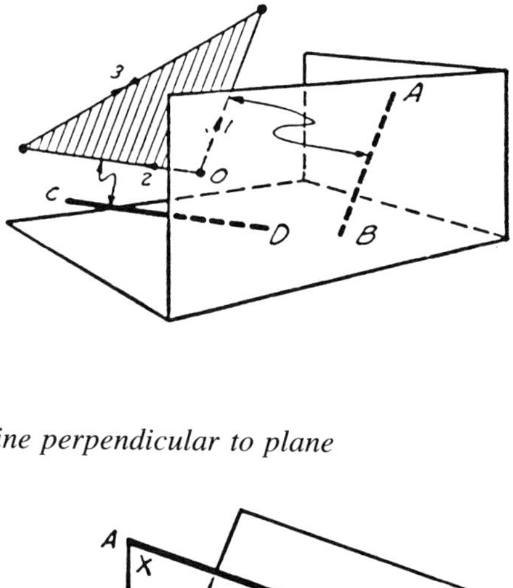

PROBLEM 13 *Line perpendicular to plane*

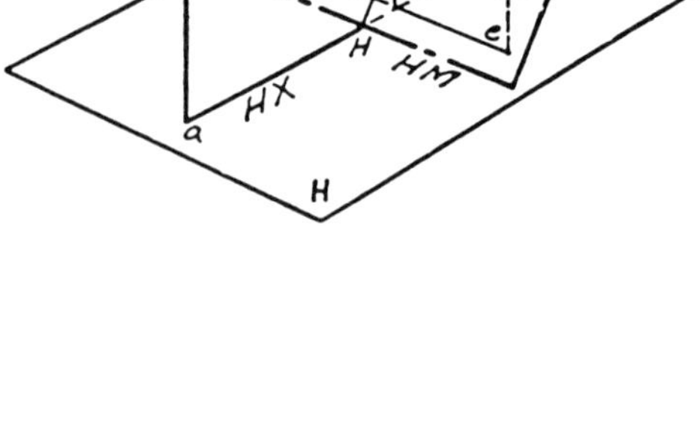

PROBLEM 14 *Plane perpendicular to plane*

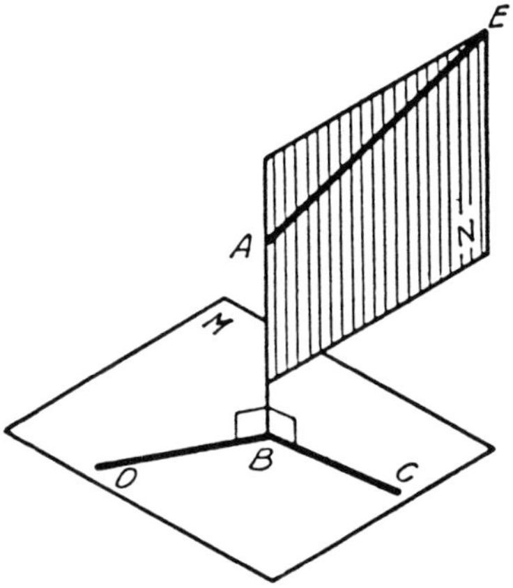

PROBLEM 15 *Intersection of lines and planes*

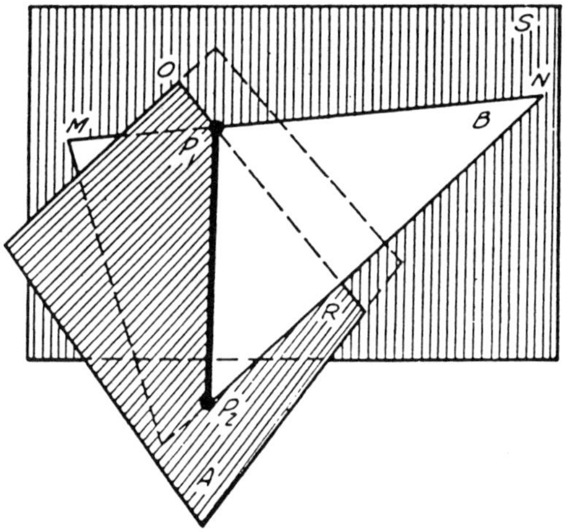

6 Advanced Point/Line/Plane Analysis

This chapter contains several CADD problem situations that most often arise during the descriptive geometry display process. The series of display problems consists of the revolution of point/line and planes. Each of the display situations is stated as a section heading; under each heading is the problem analysis, solution, and CADD construction required for the display.

If lines and planes are rotated into a position coinciding with or parallel to the coordinate planes of projection, the condition for CADD display is achieved where all properties of lines and planes are measurable. Lines may be measured for length, direction, and magnitude. Planes can be assigned areas and solids can be given volumes.

6.1 REVOLVE A POINT ABOUT A LINE

Analysis

The term *revolution* is the CADD rotation of motion in a circular path. Therefore, if a given point *O* shown in Figure 6.1 moves so that it remains in a given plane *R* and at a fixed distance from another point *C* in that plane, point *O* will describe the circumference of a circle lying in plane *R* with center at *C* and with a radius *CO*. Point *O* is said to revolve about *C* as a center.

If a line *AB* is displayed perpendicular to plane *R* through point *C*, point *O* is said to revolve about line *AB* as an axis. Point *O* is the revolving point, plane *R* is the plane of revolution, point *C* is the center of revolution, line *CO* is the radius of revolution, and line *AB* is the axis of revolution. The circle shown in Figure 6.1 is called the circle of revolution.

Solution

The problem of displaying a point revolving about a line involves only the determination of a certain definitely fixed circle. It was shown in Chapters 4 and 5 that the 3-D display of a circle involves the fixing of the three characteristics of a circle, namely, its plane, center, and radius. When these three elements are displayed, the circle and every point on its circumference are displayed. Since the plane of revolution contains the revolving point and is perpendicular to the axis of revolution, the center can be displayed also.

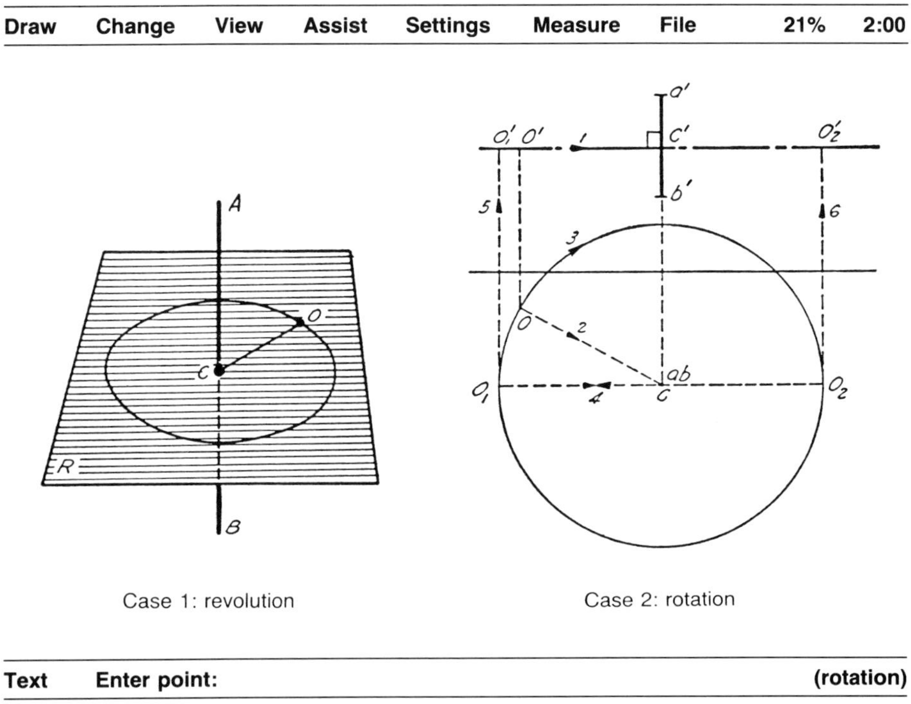

Figure 6.1 CADD problem 1.

CADD Construction

Given: Point *O* and line *AB*.

Required: Display the relationship shown in Figure 6.1.

This CADD problem called the principle of revolution has many applications in the further study of CAD/CAE descriptive geometry. The determination of values (lengths, magnitudes, or volumes) by CADD methods requires that the true distance, angle, or areas be measured. This implies that the value to be measured was displayed in the plane of measurement, or that it has been displayed upon or placed in that measurement plane in such a manner that the projected description has the same value as the original. This can only be true when the value is made parallel to the plane of measurement. This is the only CADD process no matter what name it is given: rotation, revolution, or auxiliary projection.

6.2 TRUE LENGTH OF A LINE

Analysis

The portion of a line between any two points can be displayed in its true length if the line lies in the plane of measurement or is parallel to it. A line may be rotated parallel to a plane of measurement as shown in Figure 6.2 (case 1) by revolving one extremity about a vertical line through the other extremity until they are the same distance from a reference line *(BL)*.

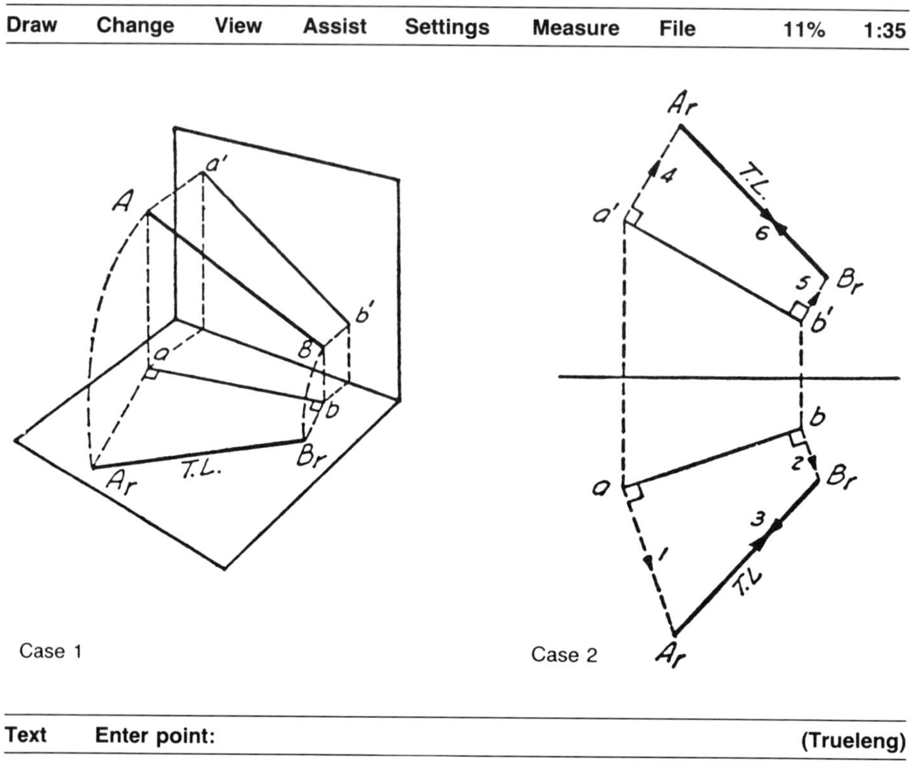

Figure 6.2 CADD problem 2.

Solution

If one of the two points defining the line in case 1 of Figure 6.2 is revolved parallel to *BL*, the line *ab* is displayed in true length as *a'b'*.

CADD Construction

Given: Line *AB*, Figure 6.2 (case 2).

Required: True length of *AB*. Begin the display by constructing the base line (step 1); label it *BL* as shown in Figure 6.2 (case 2). Next place a construction line between *b* and *b'* (step 2), which is the projection line for point *B*. Label the intersection of *BL* as *c'*. Step 3 is the rotation of line *ab* and label it *a(r)* and *bc* (step 4). Step 5 is the projection of *a(r)'* back to *BL;* then label the intersection of *BL* as *a(r)'*. Step 6 connects *a(r)'* and *b'*, which is the true length and can be labeled *TL* as shown in Figure 6.2.

6.3 DISTANCE BETWEEN A POINT AND A PLANE

Analysis

The distance from a point to a plane is measured along a perpendicular from the point to the plane. The condition of perpendicularity can be seen in Figure 6.3

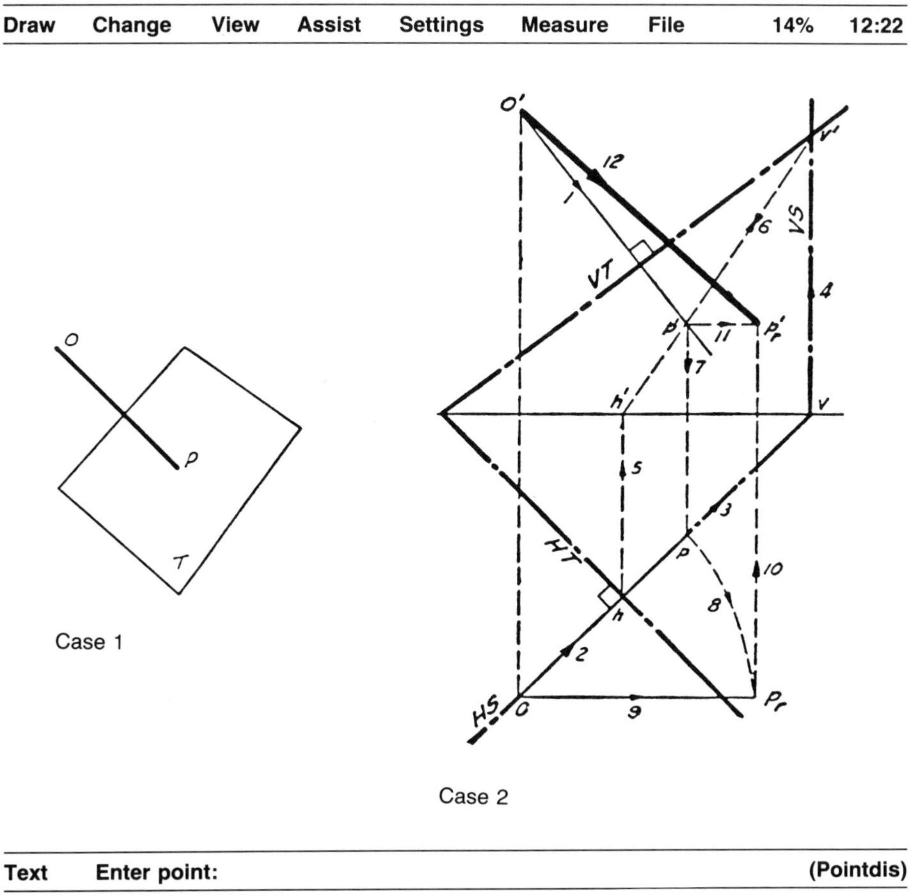

Figure 6.3 CADD problem 3.

(case 1). The distance is the length of the portion of the perpendicular from the point to the plane. This distance can be displayed as a true length line and edge view of the plane.

CADD Construction

Given: Plane T and point O, Figure 6.3.

Required: Distance from O to T. Begin the display construction with step 1, which is a perpendicular construction line through the edge view of plane T (the edge view is the vertical trace of T). Step 2 is the perpendicular construction line for the horizontal trace. Steps 3 and 4 are the intersection of the perpendiculars with BL. Step 5 is the projection of intersection h to the BL. Step 6 is the connection of h' and v'. Step 7 is the connection of p' and p. Steps 8 and 9 are the rotation of *op;* label this rotation $O,P(r)$. Step 10 is the projection of $P(r)$ to form $P'(r)$. Steps 11 and 12 are the construction of the true length line $O'P'(r)$.

6.4 DISTANCE BETWEEN A POINT AND A LINE

Analysis

The distance between a point and a line is measured along a perpendicular from the point to the line. A plane through the given point perpendicular to the line contains the perpendicular line that can be measured to determine the distance between them.

Solution

Once the plane has been established as shown in Figure 6.4 (case 1), the solution is the same as in section 6.3.

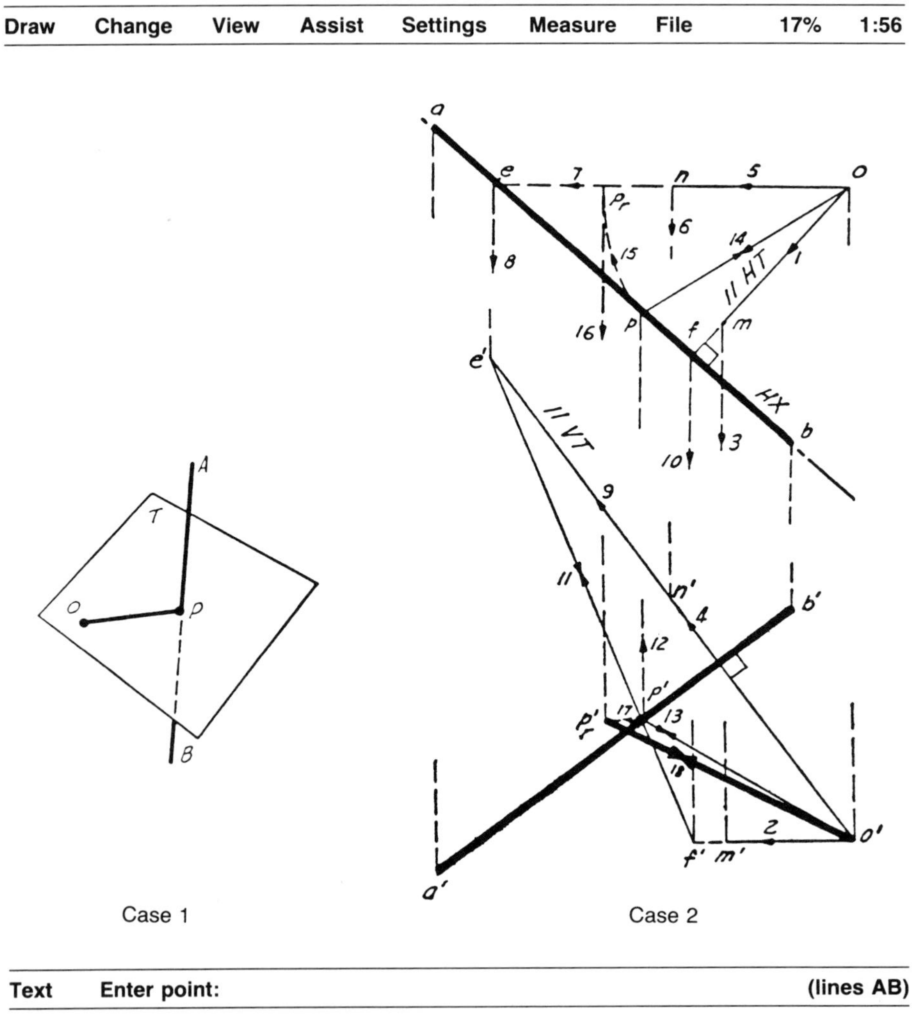

Figure 6.4 CADD problem 4.

CADD Construction

Given: Point *O* and line *AB*.

Required: Distance *O* to *AB*. Plane *T* is passed through point *O* perpendicular to line *AB* as shown in Figure 6.4 (case 2). Step 1 is the construction of line *fo;* step 2 is the projection to form *f'o'*. Steps 3 and 4 produce a parallel construction line to *f,f';* label it *m,m'*. Steps 5 and 6 are used to produce another parallel construction line; label it *n,n'*. Steps 7 and 8 are projections from *O* to line *ab* and *e',O'*. Steps 9 and 10 produce lines *e'O'* and *f'm'O'*. Steps 11 and 12 are used to display *e'f'* and *p,p'*. Steps 13 and 14 produce *p'O'* and *PO*. Steps 15 and 16 are rotation of *PO* to produce *p(r)* and *p'(R)*. Steps 17 and 18 are used to display the true distance between the point and the line.

6.5 DISTANCE BETWEEN PARALLEL LINES

Analysis

The distance between two parallel lines is measured along a common perpendicular connector. The distance is the true length of the perpendicular connector. Two cases for parallel lines are shown in Figure 6.5; case 1 assumes a plane surface forms along the perpendicular connector, and case 2 assumes a plane is formed along one of the lines *(AB)*.

Solution

A plane is used as a reference surface for either case of parallel lines. If the reference surface is displayed as an edge view, the true distance between the lines can be measured.

Draw Change View Assist Settings Measure File 19% 2:46

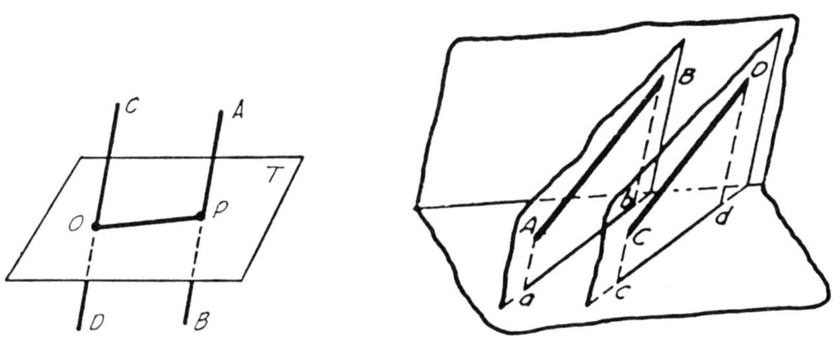

Case 1 Case 2

Text Enter point: (lines CD)

Figure 6.5 CADD problem 5.

6.6 DISTANCE BETWEEN PARALLEL PLANES
Analysis

The distance between two parallel planes is measured along a common perpendicular connector. The distance is the true length of the perpendicular connector. Two cases for parallel planes are shown in Figure 6.6. Case 1 assumes that the planes are in edge view and the perpendicular connector needs to be displayed, and in case 2 the planes are displayed in orthogonal views.

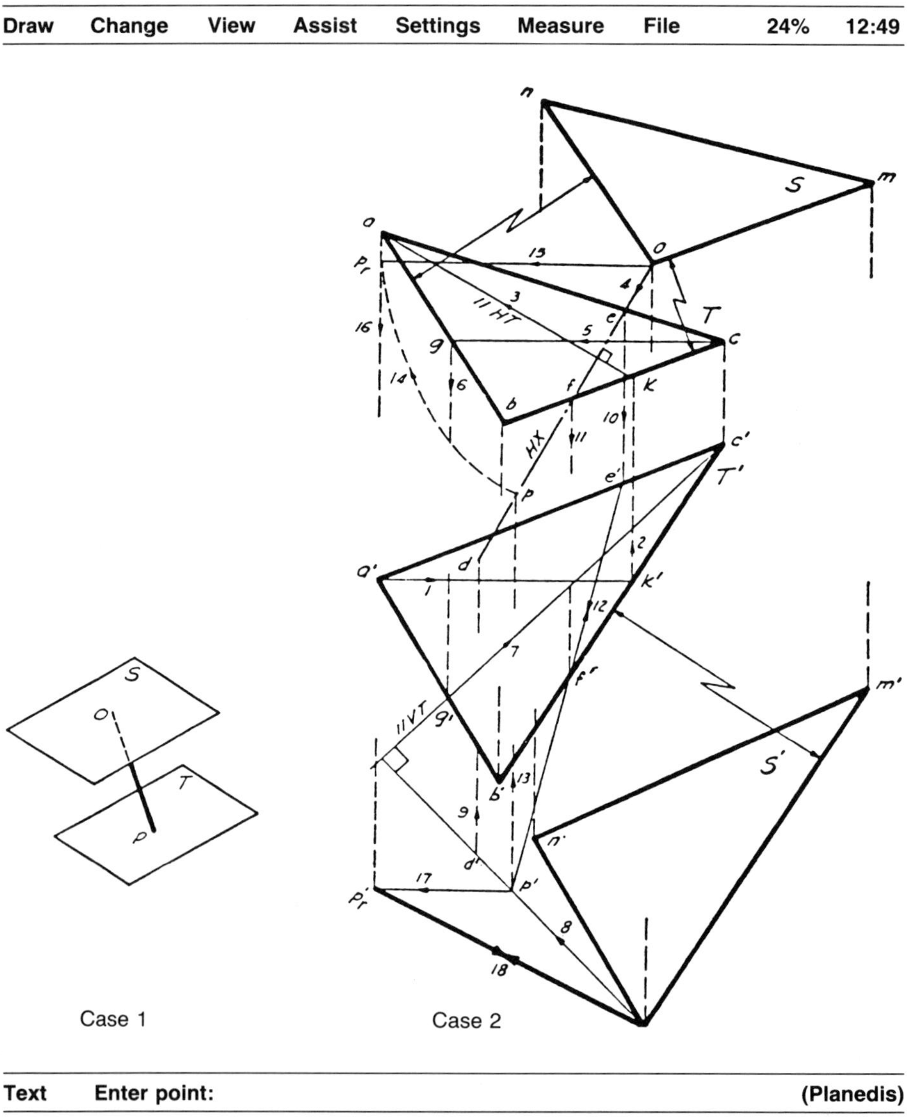

Figure 6.6 CADD problem 6.

Solution

For either case of parallel planes a line is used as a reference connector. If the reference connector is displayed as a true length, the true distance between the planes can be measured.

CADD Construction

Given: Parallel planes *S (OMN)* and *T(ABC)* shown in Figure 6.6.

Required: Distance between *S* and *T*. Begin the display construction with step 1, which is a parallel construction line *(a'k')*; project this to the opposite view (step 2). Step 3 is a *TL* (ak) parallel to the horizontal trace. Repeat this process as shown

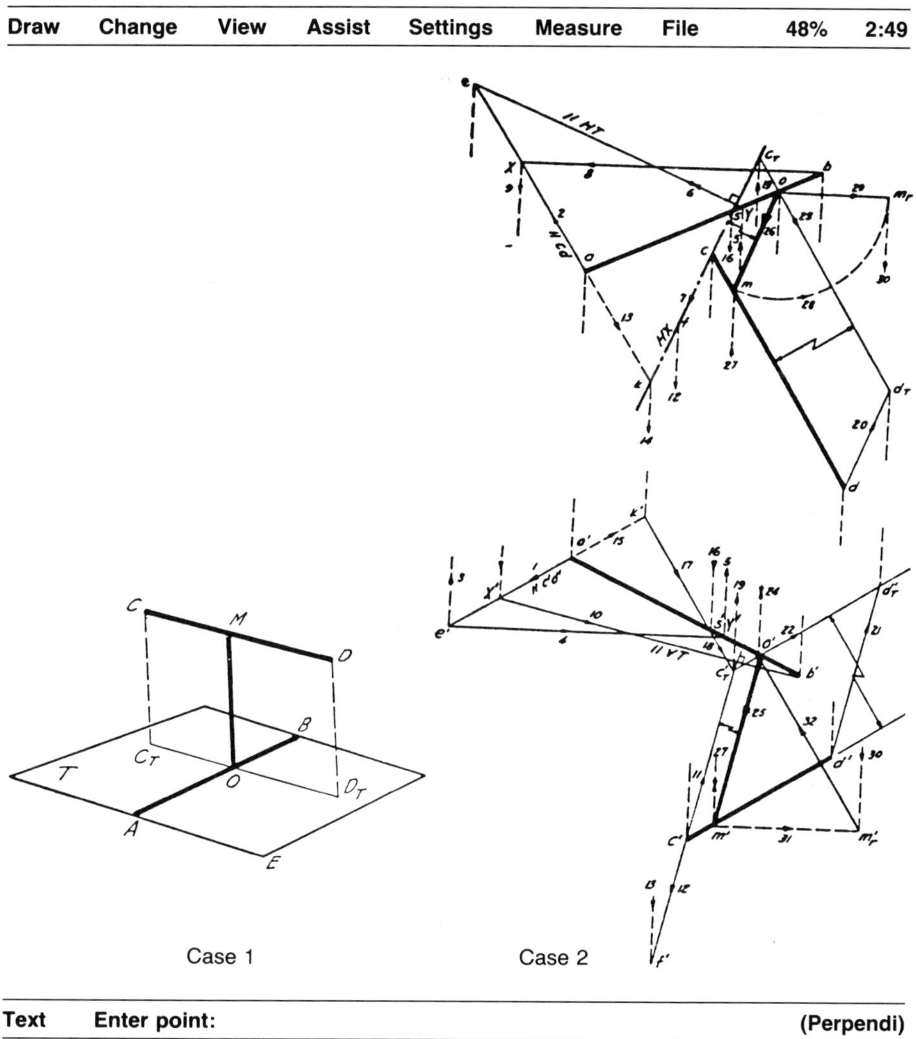

Figure 6.7 CADD problem 7: common perpendicular.

in steps 4 through 7. Next construct the perpendiculars: d' in steps 8 and 9, e' in step 10, f' in step 11, p' in step 12, and p in step 13. Now the rotations can be done as shown in steps 14 and 15 and 16 and 17 *(pr)*. The distance between the plane is shown in step 18.

6.7 COMMON PERPENDICULAR BETWEEN TWO LINES

Case 1 of Figure 6.7 is the analysis and case 2 is the construction and solution of this common CAE situation. If a line *CD* is parallel to a plane *T*, which contains line *AB*, then the common perpendicular is displayed as line *MO*.

6.8 BISECTOR OF AN ANGLE

Case 1 of Figure 6.8 is the analysis and case 2 is the construction and solution of this commonly used CAE technique. An angle can be bisected only in a view that shows its true value. The plane of the angle must therefore be revolved about any horizontal line into a horizontal display position.

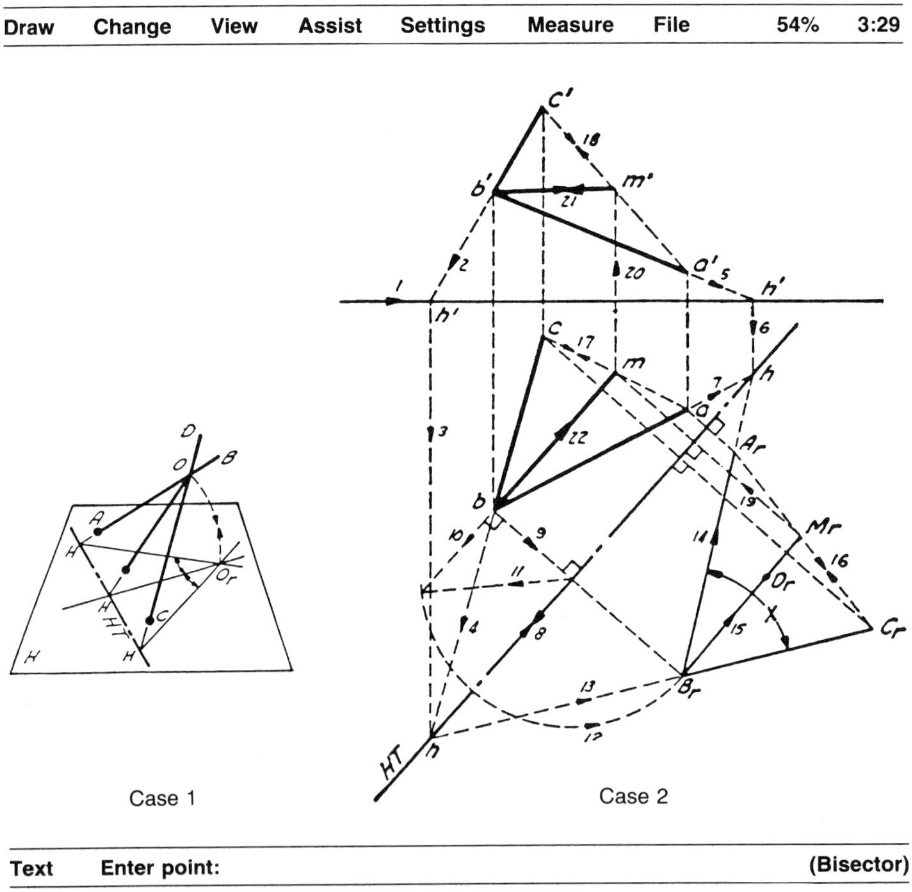

Figure 6.8 CADD problem 8: bisect an angle.

6.9 TRUE SIZE AND SHAPE OF A PLANE

Figure 6.9 is the construction and solution of the true size and shape of a plane *MNS* in the plane *AOB*. A horizontal line *AE* is chosen in plane *AOB* and revolved until it is parallel to the base line *(BL)*. The true shape and size is displayed along *BL*.

6.10 PLANE LOCATION

Figure 6.10 is the construction and solution of the true size and shape of a plane *TSMN* inserted along side the plane *AOB*. A horizontal line *AE* is used again in plane *AOB* and revolved until it is parallel to the base line *(BL)*. The true shape, size, and location is displayed along *BL*.

6.11 LINE AND PLANE LOCATION

Figure 6.11 (case 1) shows the analysis and case 2 shows the construction and solution of the insertion of a line *CO* at a specified angle with line *AB*, which is at a specified angle with plane surface *T*.

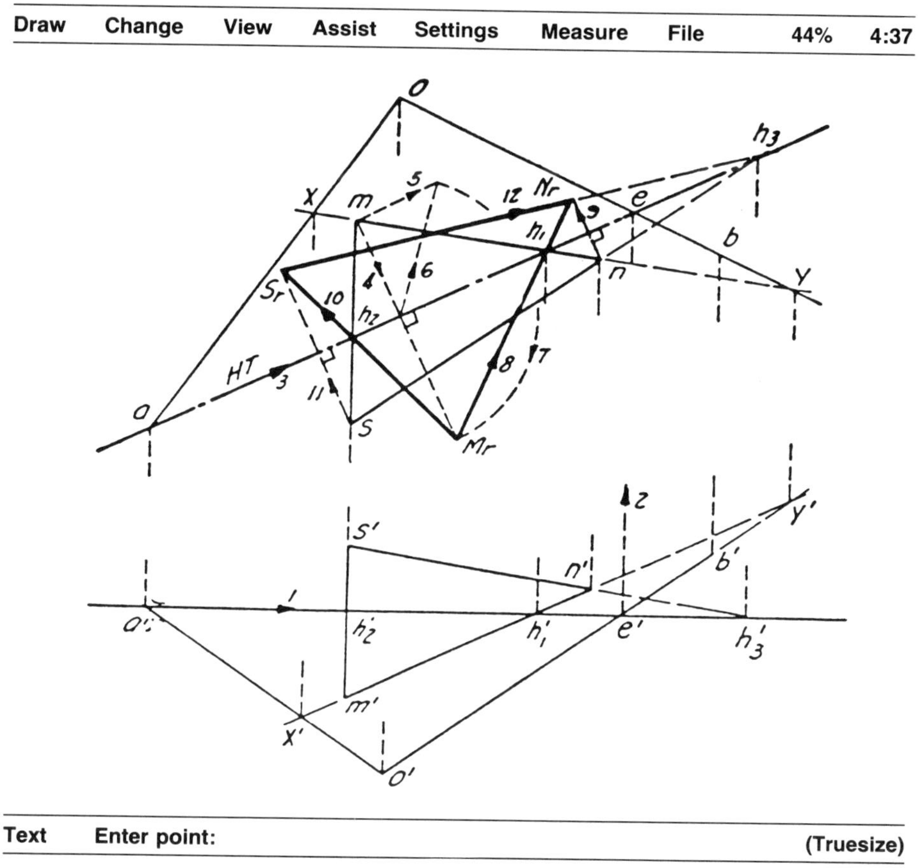

Figure 6.9 CADD problem 9: true size and shape.

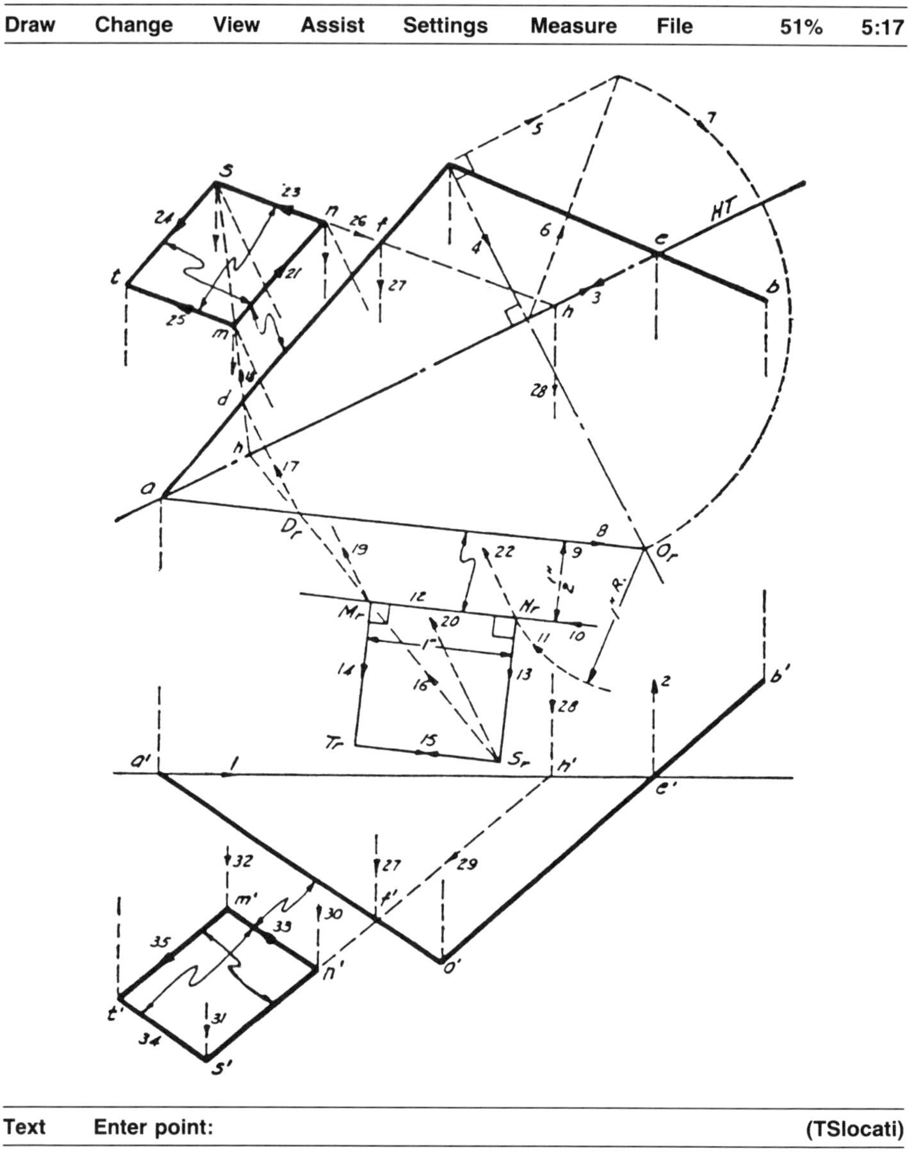

Figure 6.10 CADD problem 10: true size, shape, and location.

6.12 PLANE AND PLANE LOCATION

Case 1 of Figure 6.12 is the analysis and case 2 is the construction and solution of the insertion of a plane *S* at a specified angle (180 degrees) with plane *R*, which is at a specified angle (90 degrees) with plane surface *S*.

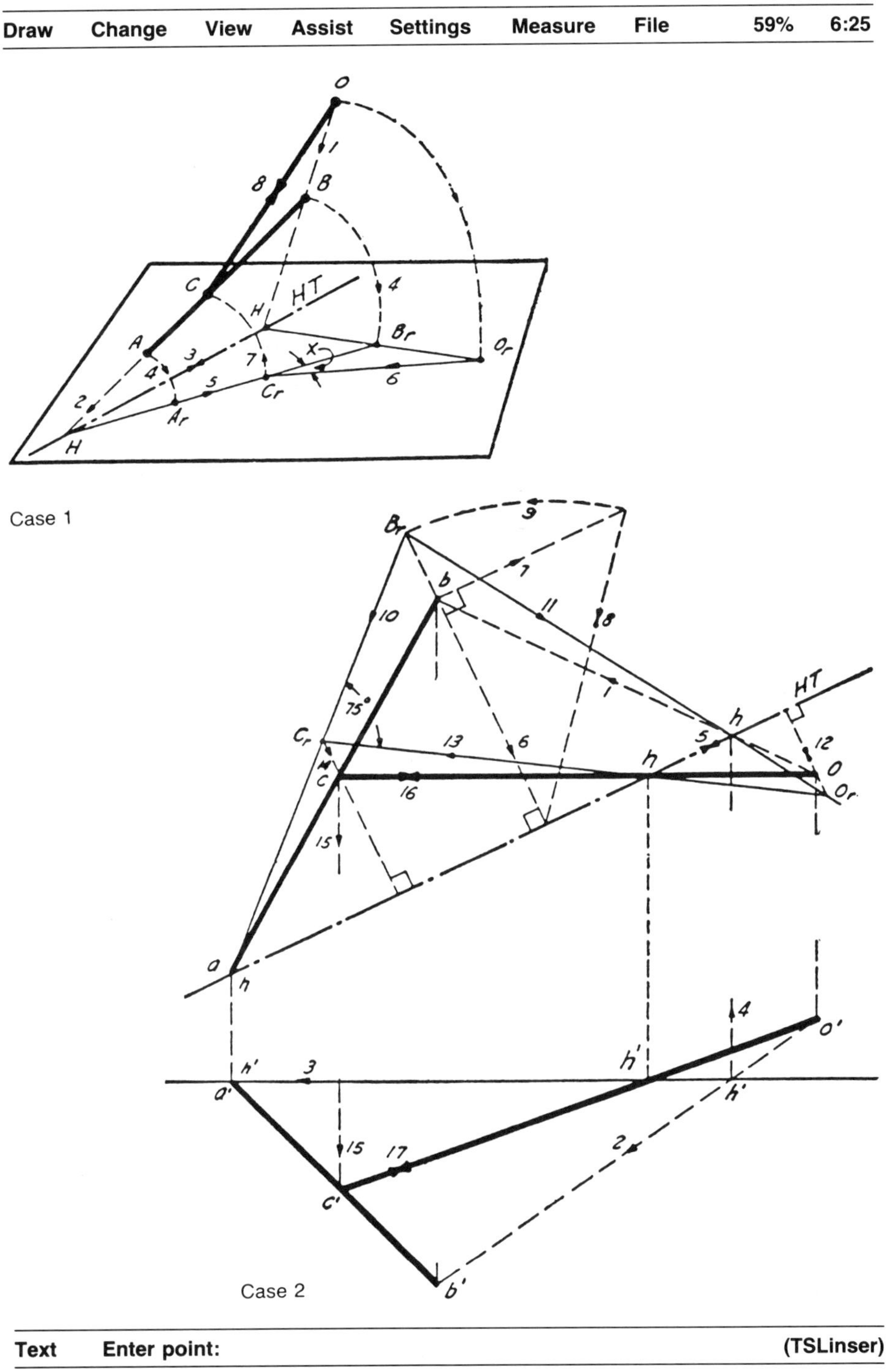

Figure 6.11 CADD problem 11: add a line into a location.

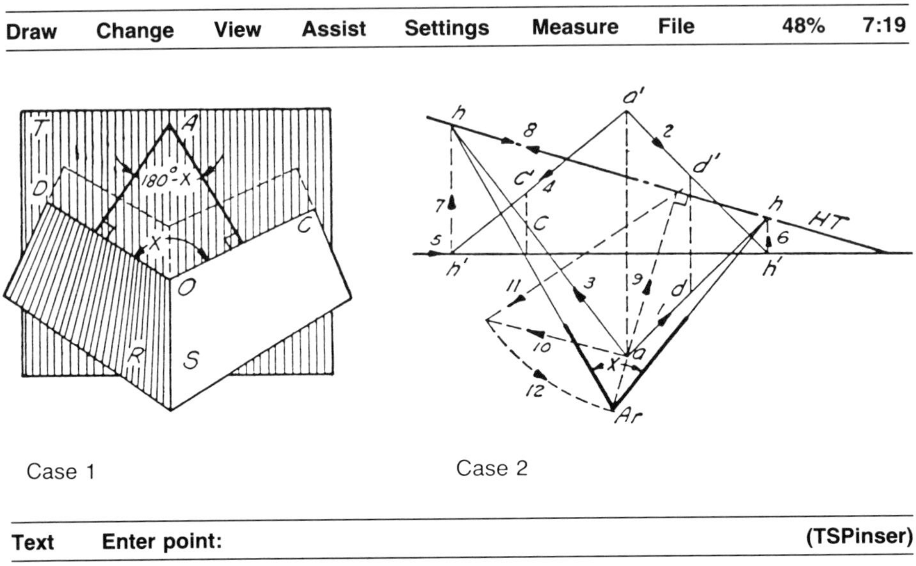

Figure 6.12 CADD problem 12: add a plane into a location.

6.13 CHAPTER PROBLEMS

The problems appearing at the end of this chapter have been selected to introduce the basic types of broad, general problems encountered by an engineer or engineering technician using a CADD display system. These problems emphasize the interdisciplinary aspects of constructive geometry, which require a number of related pieces of CADD software. The software used for the production of problems in this chapter can be ordered from Autodesk, Inc.

PROBLEM SET 1 *Revolution*

In each problem display the line defined.

1.1 Line *AB* (Fig. 6.1 [1])
1.2 Line *CO* (Fig. 6.1 [2])
1.3 Line *AB* (Fig. 6.2 [1])
1.4 Line *AB* (Fig. 6.2 [2])
1.5 Line *OP* (Fig. 6.3 [1])
1.6 Line *OP* (Fig. 6.3 [2])
1.7 Line *AB* (Fig. 6.4 [1])
1.8 Line *AB* (Fig. 6.4 [2])
1.9 Line *CD* (Fig. 6.7 [1])
1.10 Line *CD* (Fig. 6.7 [2])

PROBLEM SET 2 *Rotation*

In each problem display the plane defined.

2.1 Plane *R* (Fig. 6.1 [1])
2.2 Plane *T* (Fig. 6.3 [1])
2.3 Plane *S* (Fig. 6.5 [1])
2.4 Plane *T* (Fig. 6.5 [2])
2.5 Plane *T* (Fig. 6.6 [1])
2.6 Plane *T* (Fig. 6.7)
2.7 Plane *MNS* (Fig. 6.9)
2.8 Plane *MNST* (Fig. 6.10)
2.9 Plane *T* (Fig. 6.11)
2.10 Plane *S* (Fig. 6.12)

PROBLEM SET 3 *Measurement*

In each problem display the measurement defined.

3.1 True length of line *AB* (Fig. 6.2)
3.2 Distance *pt* to line (Fig. 6.4)
3.3 Distance *pt* to plane (Fig. 6.3)
3.4 Distance between lines (Fig. 6.5)
3.5 Distance between planes (Fig. 6.6)
3.6 Common perp distance (Fig. 6.7)
3.7 Bisector degrees (Fig. 6.8)
3.8 Area of plane (Fig. 6.9)
3.9 Plane location (Fig. 6.10)
3.10 Line/plane location (Fig. 6.11)

7 Applications for Point, Line, and Plane Measurements

This chapter contains several CADD applications for the problem situations that were shown in Chapter 6. The series of display applications consist of the revolution of point/line and planes. Each of the applications is stated as a section heading and under each heading is the CADD construction required for the display.

If lines and planes are rotated into a position coinciding with or parallel to the coordinate planes of projection as demonstrated in Chapter 6, the condition for CADD display is achieved where all properties of lines and planes are measurable by selecting another CADD function called *check*. Lines may be measured for length by *check distance*, direction by *check angle*, and planes can be measured for areas and solids can be measured for volumes with the menu item *measure* shown in Figure 7.1.

7.1 REVOLVE A POINT ABOUT A LINE

Use Figure 6.1 to review the CADD rotation of motion in a circular path. Remember, if a given point *O* shown in Figure 6.1 moves so that it remains in a given plane *R*, it is a fixed distance from point *C* in that plane.

If a line *AB* shown in Figure 7.1 (case 1) is displayed perpendicular to plane *R* through point *C*, distance *OC* can be measured on plane *R*. Line *AB* can be measured on plane *V* or *H* (case 2) if line *AB* is parallel to either *H* or *V*. If line *AB* is oblique (not parallel to *H* or *V*), it may be rotated parallel and projected to display its true length and then measured as shown in Figure 7.1.

7.2 ANGLE OF AN OBLIQUE LINE

The angle of a line *AB* in Figure 7.2 (case 1) can be measured, or the line can be displayed in its true length if the line lies in the plane of measurement or is parallel to it. A line may be rotated parallel to a plane of measurement as shown in Figure 7.2 (case 2) by revolving one extremity about a vertical line through the other extremity until they are the same distance from a reference line *(BL)*.

7.3 ANGLE BETWEEN A LINE AND *BL*

The angle between a line and *BL*, e.g., *AB* in Figure 7.3, can be measured, or the line can be displayed in its true length if the line lies on *BL;* in Figure 7.3 (case 1) it is already *TL*. A line may be rotated parallel to *BL* and measured as shown

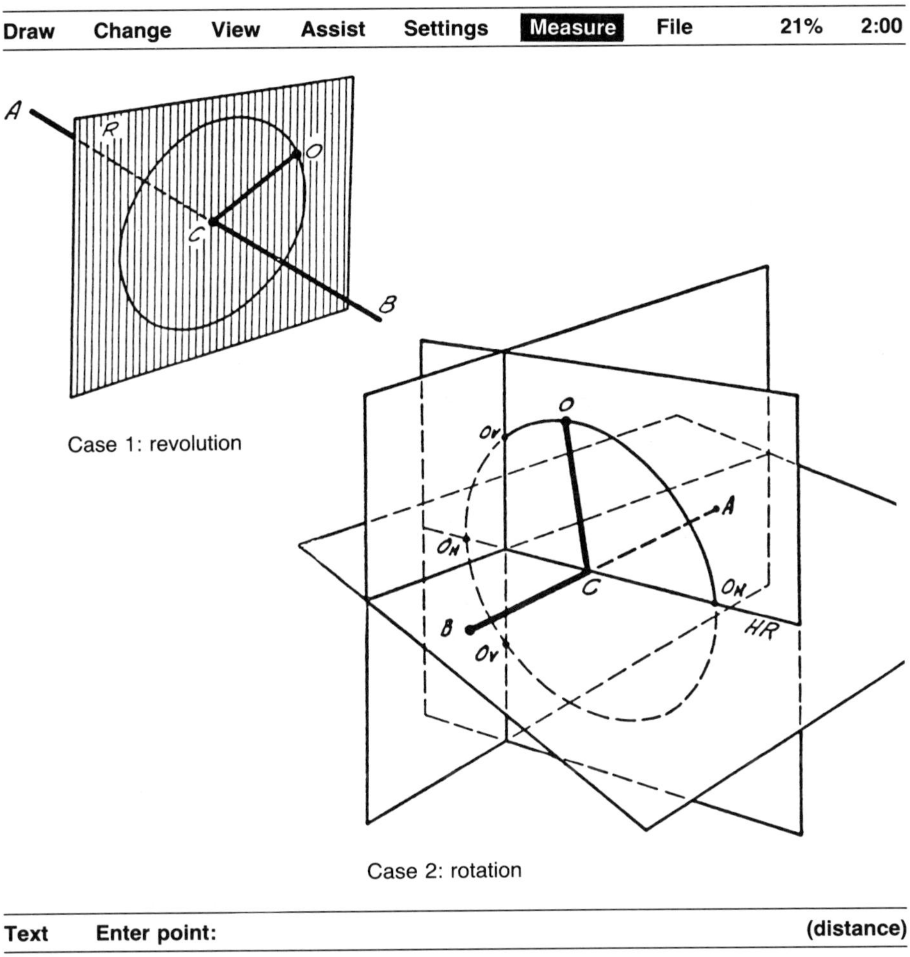

Figure 7.1 Application 1.

in case 2 by revolving one extremity about a vertical line through the other extremity until they are the same distance from *BL*.

7.4 ANGLE BETWEEN TWO LINES

The angle between two lines, e.g., *AB* and *CD* in Figure 7.4 (case 1), can be measured if the intersection point *O* is rotated into the *H* plane of measurement. Case 2 illustrates the same procedure for using *BL* to check the angle shown as angle *X*.

7.5 ANGLE BETWEEN LINE/PLANE

The angle between a line and a plane, e.g., *MN* and *T* in Figure 7.5 (case 1), can be measured if *MN* and *T* are rotated into the *H* plane of measurement.

Draw **Change** **View** **Assist** **Settings** **Measure** **File** **21%** **2:05**

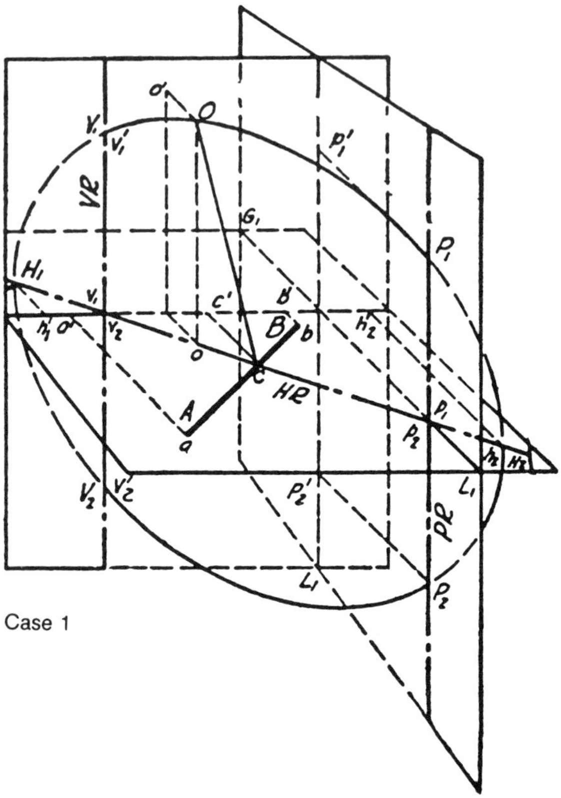

Case 1

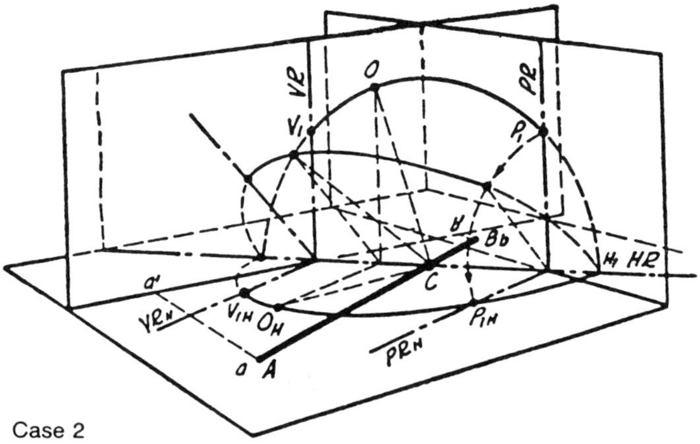

Case 2

Text **Enter point:** **(angle)**

Figure 7.2 Application 2.

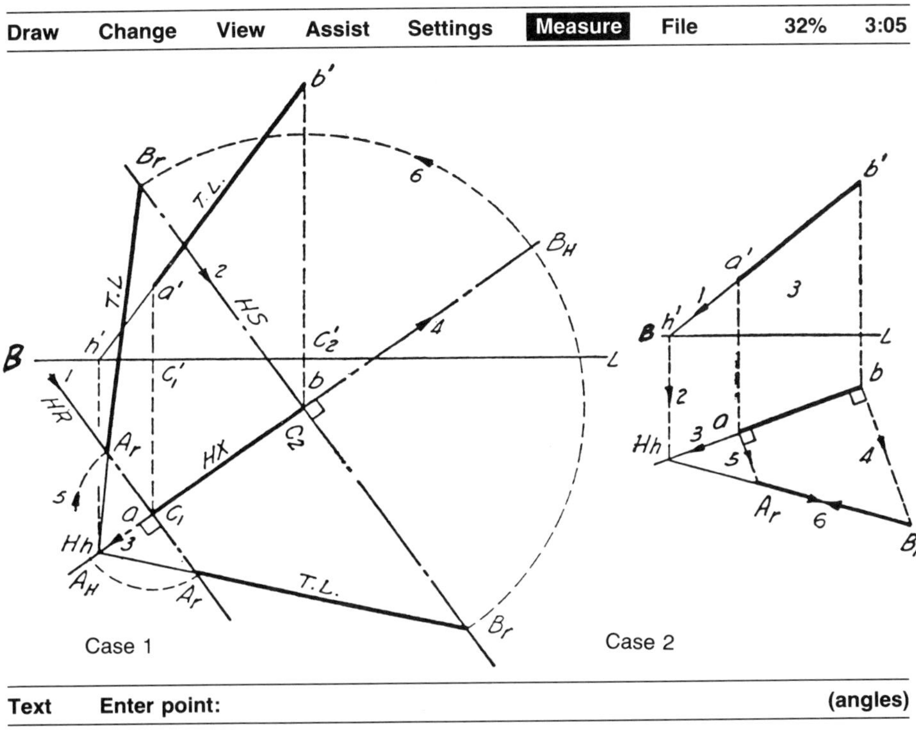

Figure 7.3 Application 3.

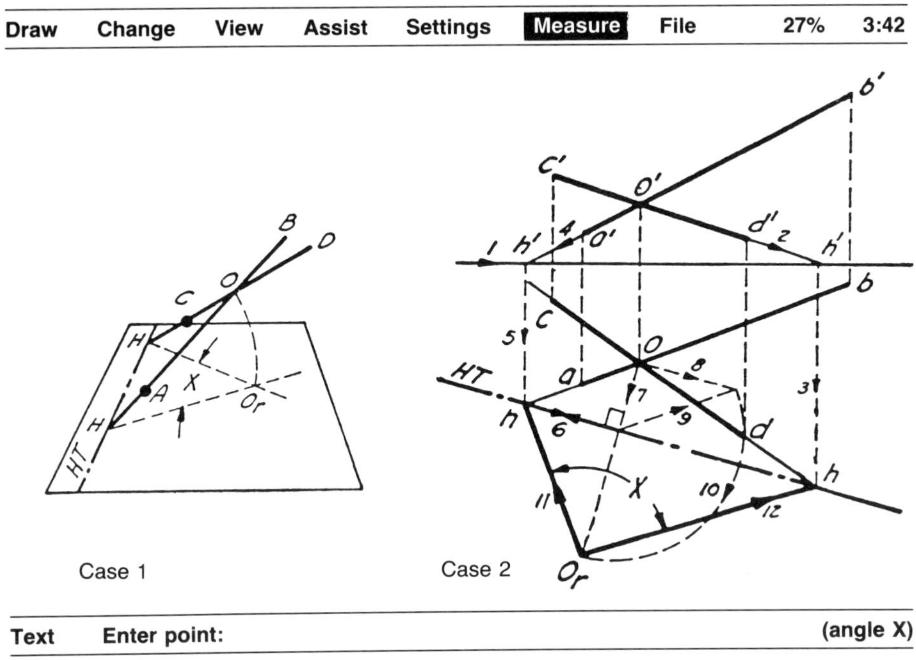

Figure 7.4 Application 4.

Draw **Change** **View** **Assist** **Settings** **Measure** **File** **42%** **4:21**

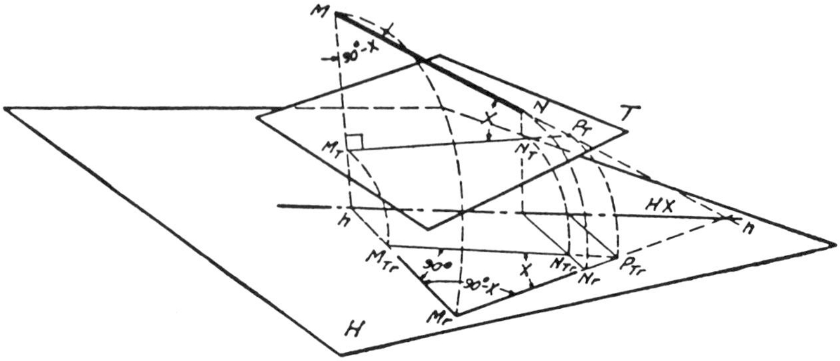

Case 1

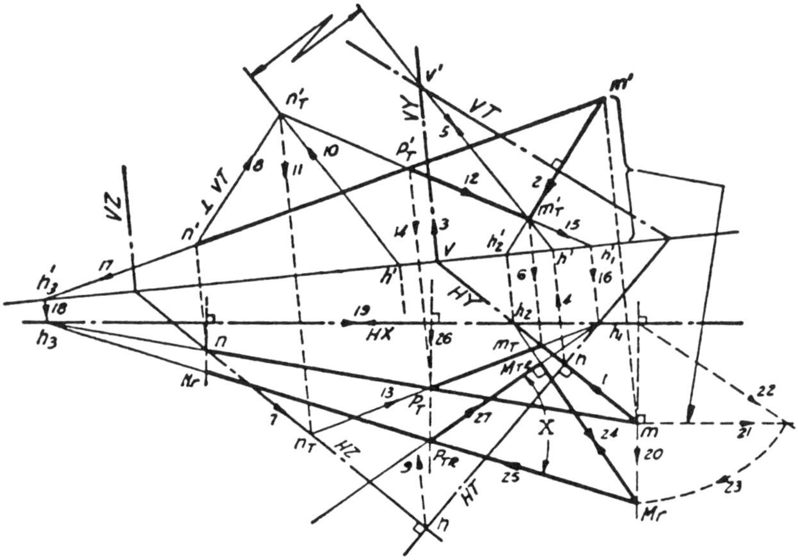

Case 2

Text **Enter point:** **(angle X)**

Figure 7.5 Application 5.

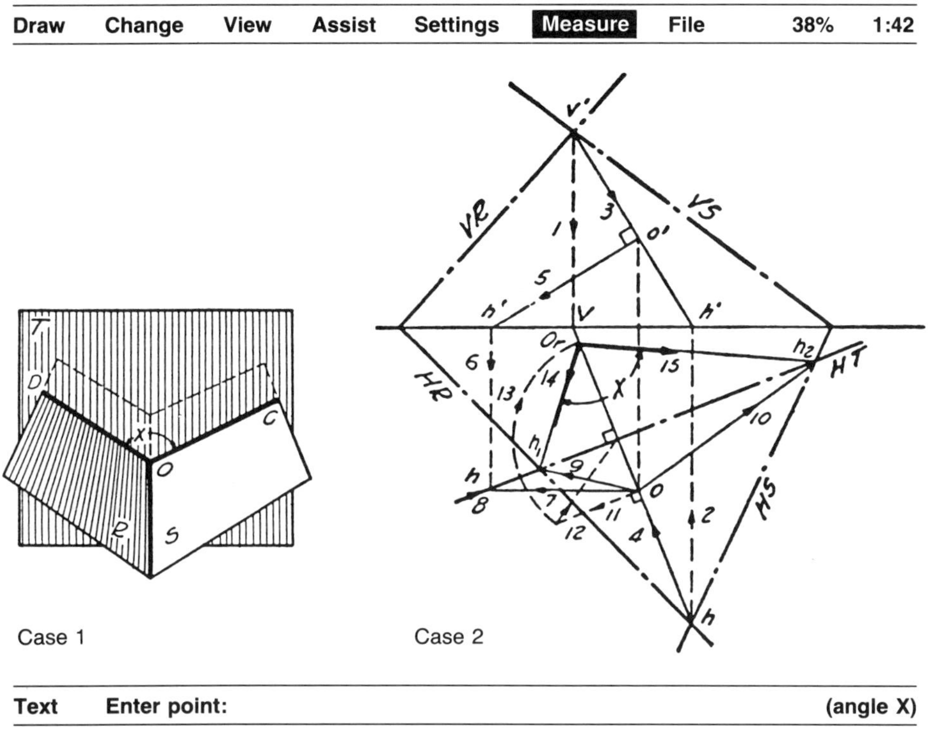

Figure 7.6 Application 6.

Case 2 illustrates the same procedure for using *BL* to check the angle shown as angle *X*.

7.6 ANGLE BETWEEN TWO PLANES

The angle between two planes, e.g., *R* and *S* in Figure 7.6 (case 1), can be measured if *R* and *S* are rotated into the *T* plane of measurement. Case 2 illustrates the same procedure for using *BL* and the edge views of *R* and *S* to check the angle shown as angle *X*.

7.7 AREA OF A PLANE

The area of a plane surface *T* bounded by points *C, O,* and *M* shown in Figure 7.7 can be measured if *O* and *M* are rotated into the plane of measurement. Three true length line segments form the boundaries of plane *T,* and each of the plane surface angles are true and can be measured.

7.8 VOLUME OF A SOLID

The volume of a solid bounded by planes *R, S,* and *T* shown in Figure 7.8 (case 1) can be measured if the areas of *R* and *S* are known and the area for the por-

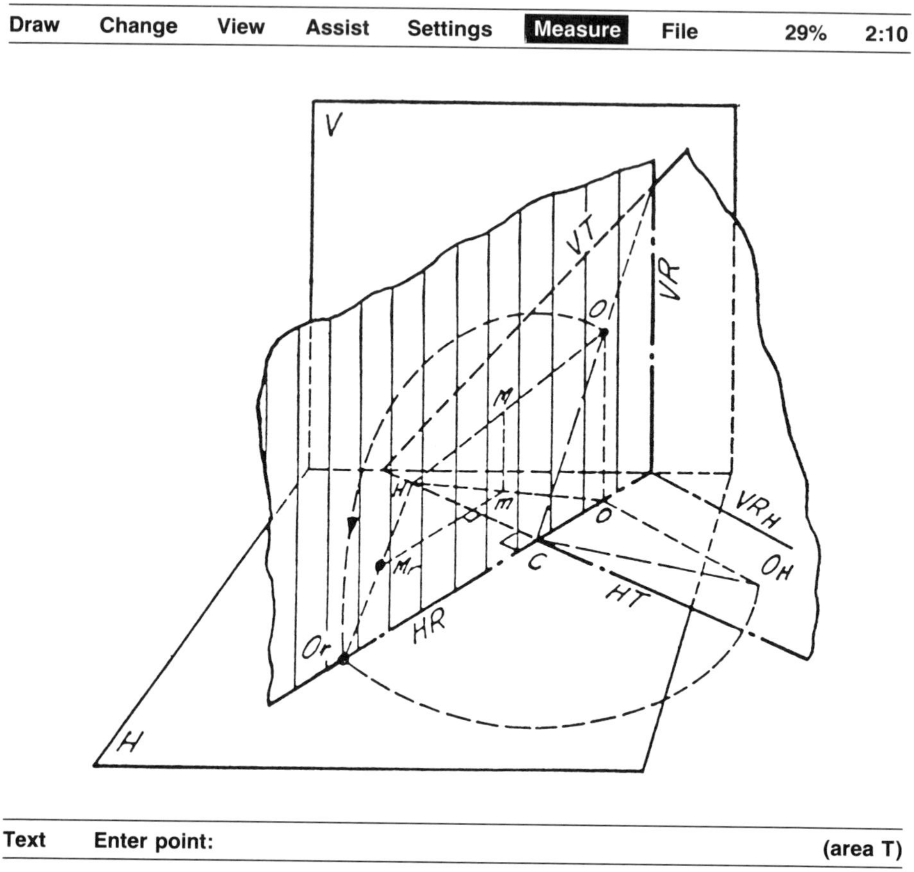

Figure 7.7 Application 7.

tion of surface *T* bounded by points *D, O, C,* and *A*. Once the area of surface *DOCA* is found, then the volume for the solid can be displayed. It is common practice to display the traces of the plane surfaces opposite the *BL* as shown in case 2.

7.9 DISPLAY SIMPLE SOLIDS

A simple solid is bounded by plane surfaces. The number of surfaces determines the display technique. Solids containing only a single surface are known as *spheres*. Two methods are used to display spheres: the meridian method shown in Figure 7.9 and the zone method shown in Figure 7.10. Both of these display methods consist of breaking up the single surface into small portions and assuming each portion to be a part of the single surface of the sphere.

The meridian method shown in Figure 7.9 consists of a series of plane surfaces spaced an equal distance, or angle, apart around the sphere. This is known in practice as the orange-peel method. The zone method shown in Figure 7.10 consists

Draw **Change** **View** **Assist** **Settings** **Measure** **File** **67%** **3:40**

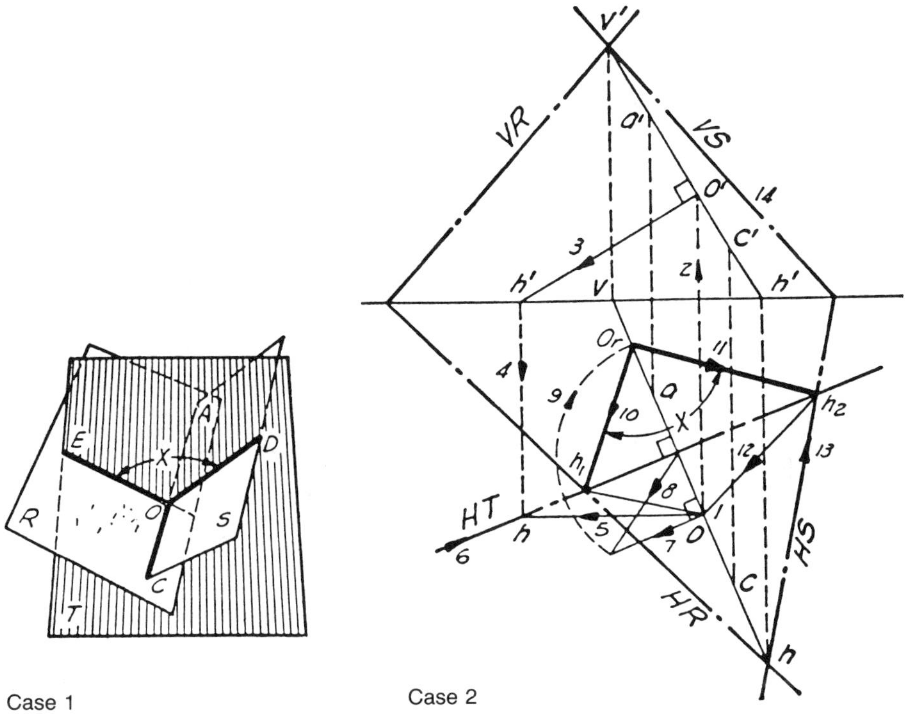

Text **Enter point:** **(volume A)**

Figure 7.8 Application 8.

Draw **Change** **View** **Assist** **Settings** **Measure** **File** **22%** **2:45**

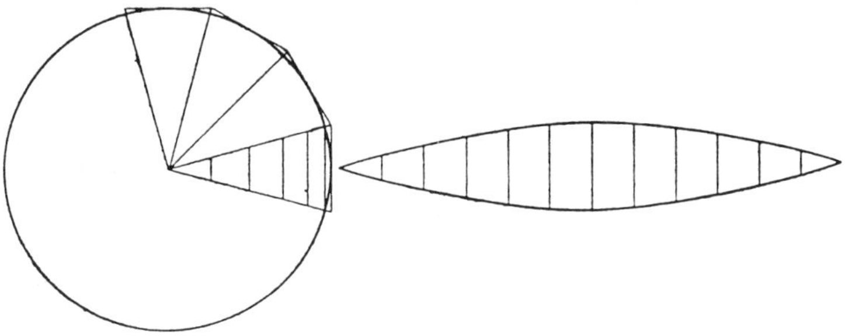

Text **Enter point:** **(spheres)**

Figure 7.9 Meridian method of display for spheres.

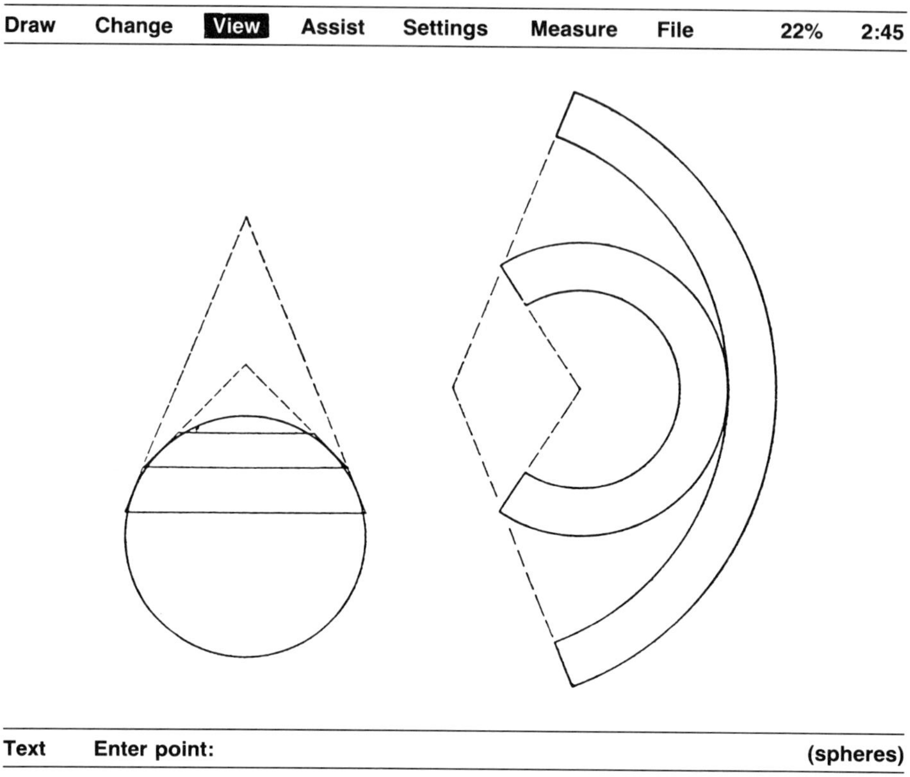

Figure 7.10 Zone method of display for spheres.

of cutting the sphere by a series of parallel planes and assuming that each section of the surface is a portion of a cone or revolution. Each of the sections has to be developed independently, for they all lie on cones of different size, that is, for half the sphere. The halves of the sphere are the same, and the sections can be displayed in reverse order for the second half.

7.10 TWO-SURFACE SOLIDS

Solids formed from two surfaces are known as cones. Figure 7.11 is the display of three types of cones. Type 1 is called *oblique*. On this cone exactly half the elements are visible in the front view and the base is parallel to *BL*. Type 2 is called a *right cone* because all elements are visible in the front and top views. Also the altitude intersects the base at right angles. Type 3 is called *elliptical-shaped* because the base appears as an ellipse in both views.

7.11 THREE-SURFACE SOLIDS

Solids formed from three surfaces are known as cylinders. Figure 7.12 is the display of a right cylinder. Two base surfaces are used along with a single side surface

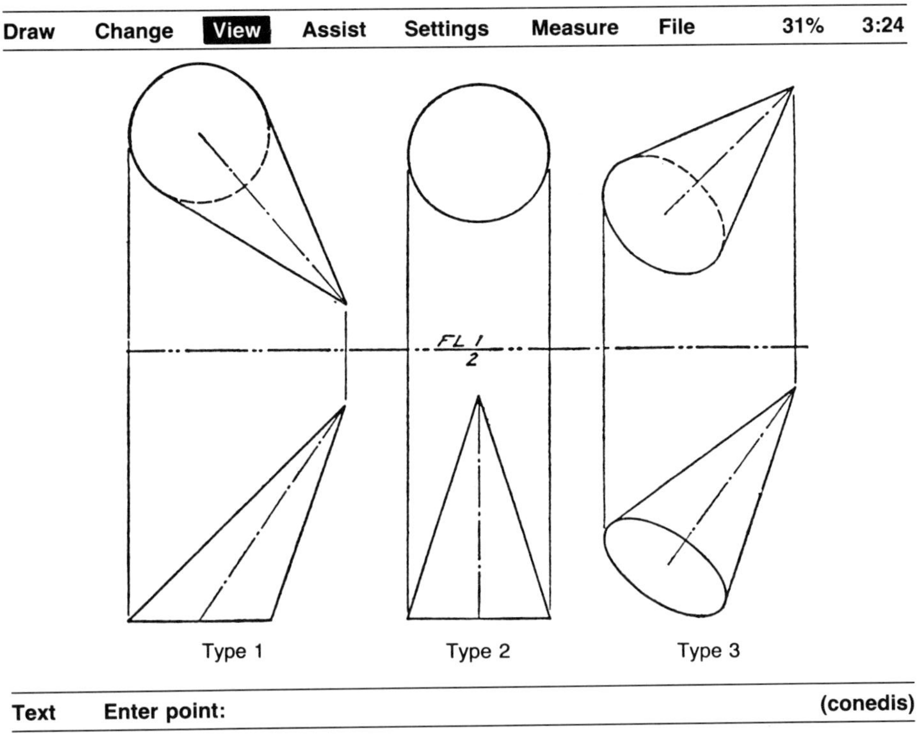

Figure 7.11 Cone displays.

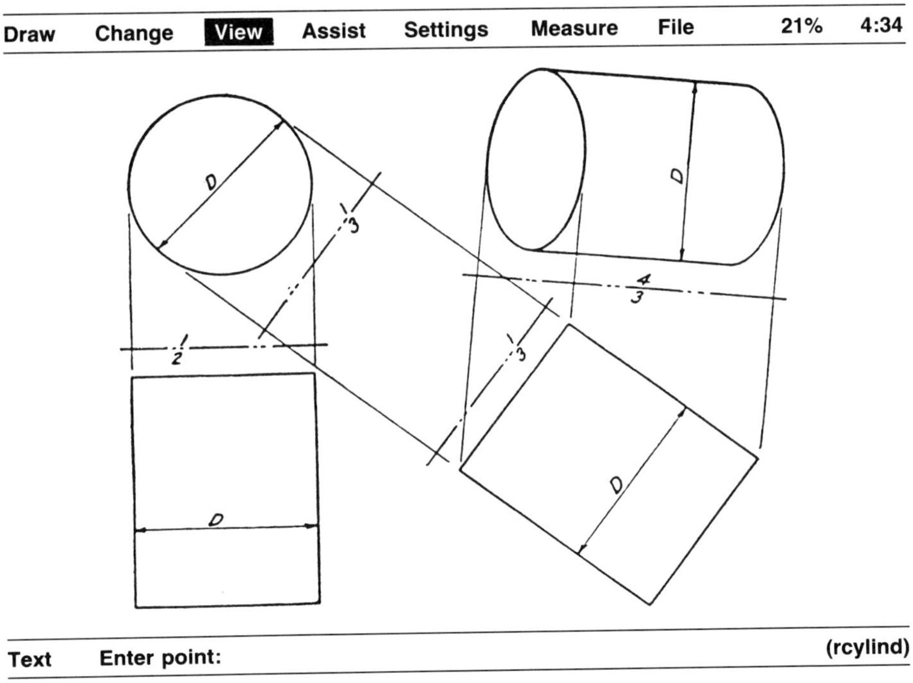

Figure 7.12 Right cylinder.

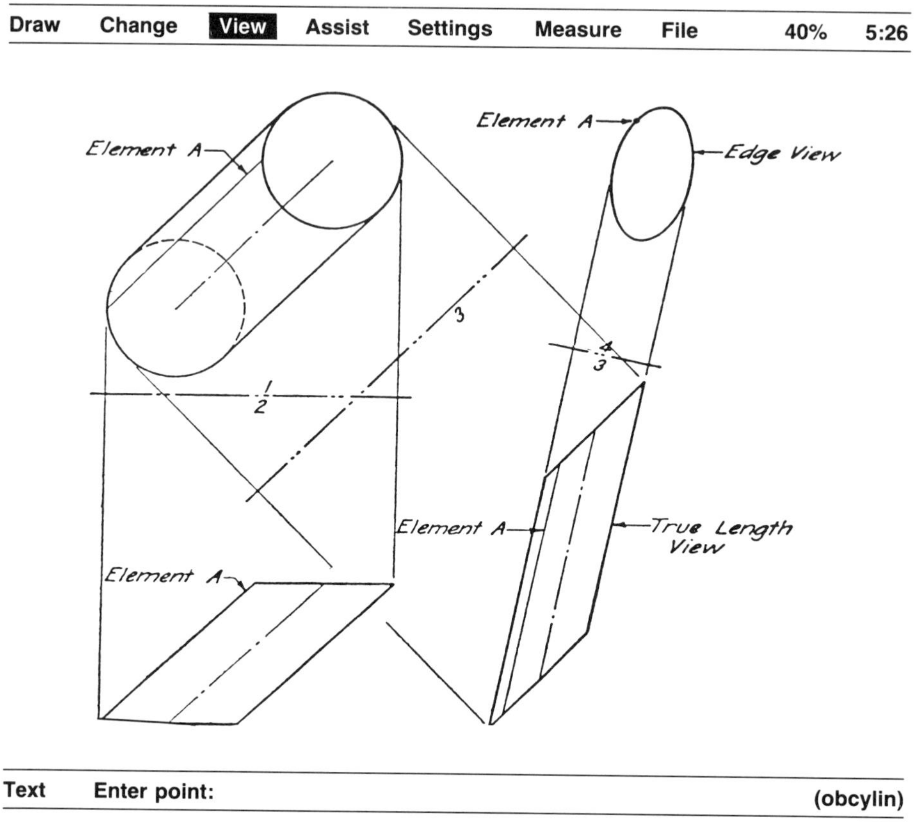

Figure 7.13 Oblique-elliptical cylinder.

wrapped around a circle *(D)*. Three types of cylinders can be displayed: right, oblique, and elliptical. Figure 7.13 is a working display of an oblique-elliptical cylinder.

7.12 MULTI-SURFACE SOLIDS

Solids formed from four or more surfaces are known as *polyhedrons*. Examples of these are prisms, pyramids, and cubes. The solid object can usually be displayed in its simplest position in the view showing the plane surfaces as edges or in true shape. Whether it's possible or not depends upon the polyhedron chosen.

7.13 DISPLAY SOLIDS ON PLANES

A solid object can usually be displayed in its simplest position in the view showing the plane as an edge. In some cases a solid might have to be stored as a working file (filename.dxf) showing the true size of the plane before the solid can be

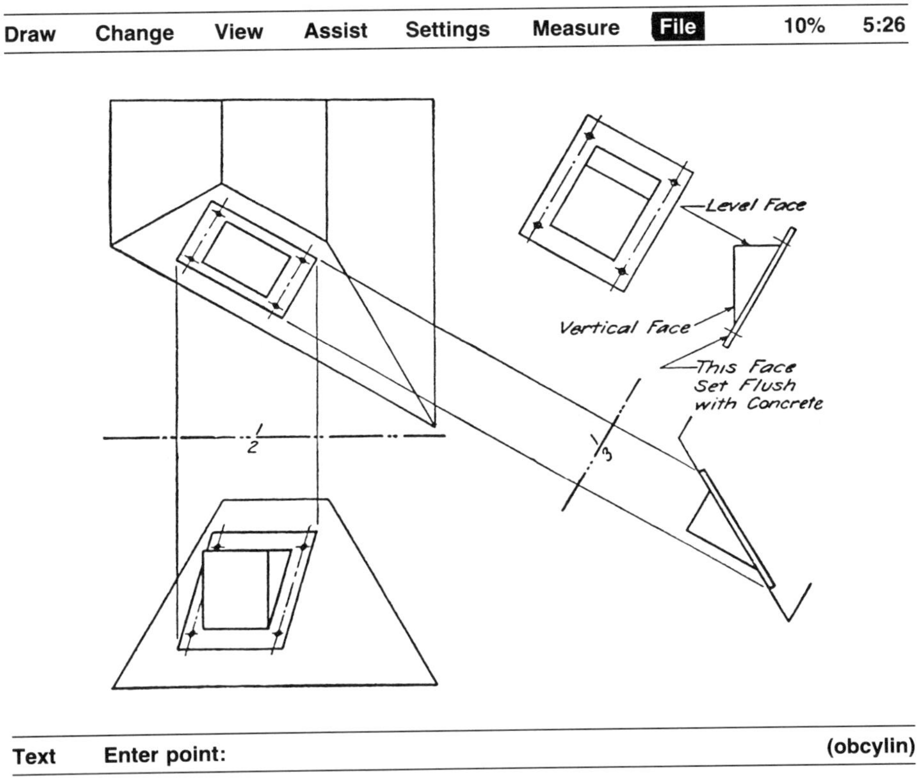

Figure 7.14 Solid and plane from dxf file.

displayed in its correct position. See Figure 7.14 for an example of a solid resting on an oblique plane.

7.14 DISPLAY PLANES ON PLANES

A plane can also be displayed in its simplest position in the view showing the plane as an edge. In some cases the plane to be placed might have to be stored as a working file (filename.dxf) showing the true size and shape of the plane before the receiving plane can be displayed in its correct position. See Figure 7.15 for an example of a circle resting on an oblique plane.

7.15 CHAPTER PROBLEMS

The problems appearing at the end of this chapter have been selected to introduce the basic types of broad, general problems encountered by an engineer or engineering technician using a CADD display system. These problems emphasize the interdisciplinary aspects of constructive geometry, which require a number of related pieces of CADD software. The software used for the production of problems in this chapter can be ordered from Autodesk, Inc.

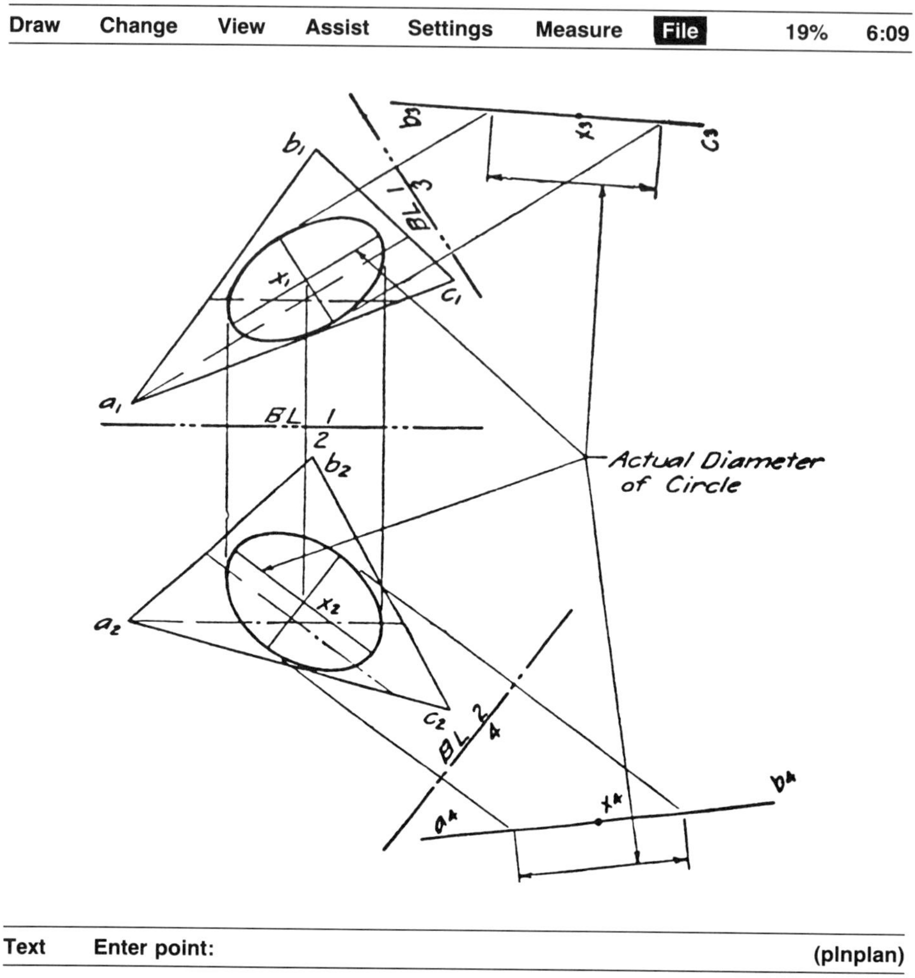

Figure 7.15 Circle and plane from dxf file.

PROBLEM SET 1 *Revolution*

In each problem display the line defined.

1.1 Line *AB* (Fig. 7.1 [1])
1.2 Line *BC* (Fig. 7.1 [2])
1.3 Line *OC* (Fig. 7.1 [2])
1.4 Line *AB* (Fig. 7.2 [1])
1.5 Line *AC* (Fig. 7.2 [2])
1.6 Line *AB* (Fig. 7.3 [1])
1.7 Line *TL* (Fig. 7.3 [2])
1.8 Line *CD* (Fig. 7.4 [1])
1.9 Line *AB* (Fig. 7.4 [1])
1.10 Line *CD* (Fig. 7.4 [2])

PROBLEM SET 2 *Rotation*

In each problem display the plane defined.

2.1 Plane *R* (Fig. 7.1 [1])
2.2 Plane *N* (Fig. 7.5 [1])
2.3 Plane *T* (Fig. 7.6 [1])
2.4 Plane *R* (Fig. 7.6 [1])
2.5 Plane *S* (Fig. 7.6 [1])
2.6 Plane *T* (Fig. 7.7)
2.7 Plane *T* (Fig. 7.8 [1])
2.8 Plane *S* (Fig. 7.8 [1])
2.9 Plane *R* (Fig. 7.8 [1])
2.10 Plane *X* (Fig. 7.15)

PROBLEM SET 3 *Measurement*

In each problem display the measurement defined.

3.1 Length of line *AB* (Fig. 7.2)
3.2 Angle of line *AB* (Fig. 7.3)
3.3 Angle between lines (Fig. 7.4)
3.4 Angle between line and plane (Fig. 7.5)
3.5 Angle between two planes (Fig. 7.6)
3.6 Area of plane (Fig. 7.7)
3.7 Volume of solid (Fig. 7.8)
3.8 Surface area of sphere (Fig. 7.9)
3.9 Volume of sphere (Fig. 7.10)
3.10 Volume of cone (Fig. 7.11)

PROBLEM SET 4 *Solids*

In each problem display the solid defined.

4.1 Solid (Fig. 7.8)
4.2 Sphere (Fig. 7.9)
4.3 Sphere (Fig. 7.10)
4.4 Cone type 1 (Fig. 7.11)
4.5 Cone type 2 (Fig. 7.11)
4.6 Cone type 3 (Fig. 7.11)
4.7 Cylinder (Fig. 7.12)
4.8 Cylinder (Fig. 7.13)
4.9 DXF (Fig. 7.14)
4.10 DXF (Fig. 7.15)

8 Planes and Solids

This chapter introduces the CADD relationship between the plane and the solid. Prior to this chapter the relationship has been the point/line/plane. The series of display solids shown in this chapter consist of the linking of planes to form solids. Each of the display situations is a section heading; under each heading is the CADD construction required for the display.

If planes and solids are built from straight lines (motion of a point in a fixed direction) as defined in Chapter 2, then this chapter would be rather useless. As pointed out in Chapter 4, however, a curved line is generated by a point moving in a constantly changing direction. Therefore, planes can have a single curvature composed of straight line segments or a double curvature composed of curved line segments.

8.1 PLANE CLASSIFICATION

Use Table 8.1 to review the CADD classification of plane surface composition. Planes are divided into two groups: straight and curved. Curved planes are divided into two groups: single and double. Planes can be displayed from the following databases: (1) conics, (2) spirals, or (3) math functions. Conics are composed of (1) circle, (2) ellipse, (3) parabola, or (4) hyperbola. Spirals are composed of (1) archimedean, (2) logarithmic, (3) hyperbolic, or (4) involutes. Mathematical functions such as sinusoid, tangent, or secant can be used, or the database may consist of trochoids or double curved surfaces as in Chapter 4.

8.2 SOLID CLASSIFICATION

Use Table 8.2 to review the CADD classification of solid composition. Solids are divided into two groups: wireform and geometric. Wireforms are either single (cylinders and cones) or warped. Warped solids can be displayed from the following databases: (1) hyperbolic paraboloid, (2) hyperboloid of revolution, (3) helicoid, (4) conoid, or (5) general directrices. Geometric solids are considered to be either a sphere, ellipsoid, paraboloid, hyperboloid, or torus. Geometric solids can be displayed from the following databases: (1) sweeps, (2) two axis equal, (3) general, (4) three axis unequal, or (5) elliptical.

Table 8.1 CLASSIFICATION OF PLANES

- Planes
  - Straight
  - Curved
    - Single
      - Conics
        - Circle
        - Ellipse
        - Parabola
        - Hyperbola
      - Spirals
        - Archimedean
        - Logarithmic
        - Hyperbolic
        - Involutes
      - Mathematical
        - Sinusoid
        - Tangent
        - Secant
      - Trochoids
        - Cycloid
        - Epicycloid
        - Hypocycloid
    - Double
      - Helices
        - Cylindrical
        - Conical
        - Spherical

Table 8.2 CLASSIFICATION OF SOLIDS

- Solids
  - Wireform
    - Single
      - Cylinders
      - Cones
    - Warped
      - Hyperbolic paraboloid
      - Hyperboloid of revolution
      - Helicoid
      - Conoid
      - General directrices
  - Geometric
    - Sphere
      - Sweeps
    - Ellipsoid
      - Two axis equal
      - General
      - Three axis unequal
    - Paraboloid
      - Sweeps
      - Elliptical
    - Hyperboloid
      - Sweep
      - Elliptical
    - Torus

Draw Change View Assist Settings Measure File 19% 10:49

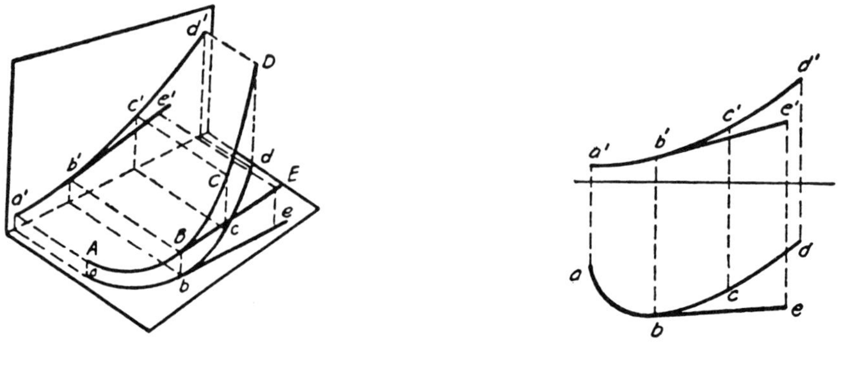

Case 1 Case 2

Text Enter point: (Planefig)

Figure 8.1 Plane projection.

Draw Change View Assist Settings Measure File 28% 11:52

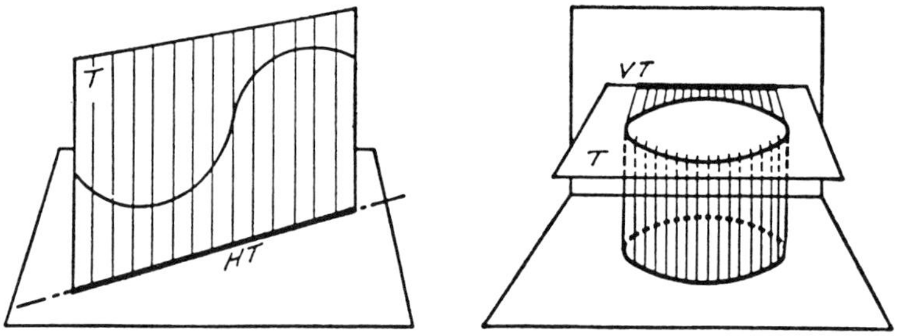

Case 1 Case 2

Text Enter point: (curvepln)

Figure 8.2 Plane projection.

8.3 PROJECTION OF PLANES

The projection of a plane is composed of the projections of the lines contained within the plane. If a sufficient number of lines are projected upon the horizontal picture plane, a surface is displayed as shown in Figure 8.1. Case 1 is the coordinate model for a plane surface and case 2 is the orthographic relationship.

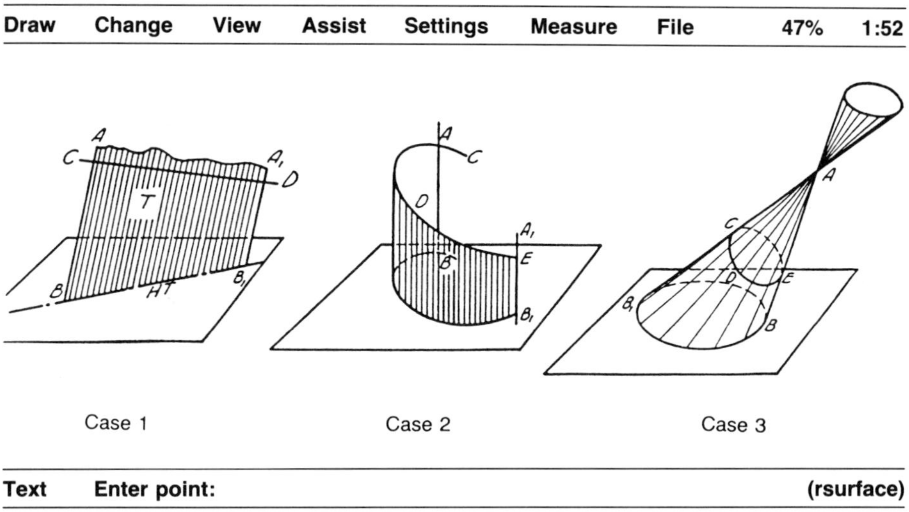

Figure 8.3 Ruled surfaces.

Plane surfaces in general are conceived as being generated by a line moving according to some law. They are called *ruled surfaces* if they can be displayed by a straight line and *curved surfaces* if the moving line is a curve.

The moving line that generates the plane surface in Figure 8.2 (case 1) is called the *generatrix*. Any position of the generatrix is called an *element* of the plane surface. The surface shown in Figure 8.2 (case 1) is called a *ruled surface*. Ruled surfaces are displayed by a straight line moving according to the equation for a circle. Ruled surfaces can also be displayed as shown in Figure 8.3. In case 1 plane surface T is displayed by a straight line moving so as to touch another straight line and having all of its positions parallel. In case 2 a given curve is used instead of the straight line, and in case 3 a single curved surface is displayed by a straight line moving so that it touches given-sized circle elements. Single curved surfaces may be displayed as cylinders, cones, or convolutes (all positions tangent to a given space curve).

8.4 CYLINDERS

A cylinder may be displayed by generating a straight line moving so that it is always parallel to its first position and constantly touching a given circle or curve as shown in Figure 8.4. If the cylinder is intersected by a plane as in case 1, the curve of intersection is a circle and is called a right cylinder. In case 2, an oblique cylinder is formed. The curves of intersections are called the *base*.

If we wish to display a definite portion of the surface forming the cylinder, a second plane may be passed parallel to the first, intersecting the cylinder in a second base. The first intersection is called the *lower base*, and the second is called the *upper base*. According to the direction of the CADD display axis, such bases might be called *front* and *rear* or *right* and *left*. The portion between the bases is

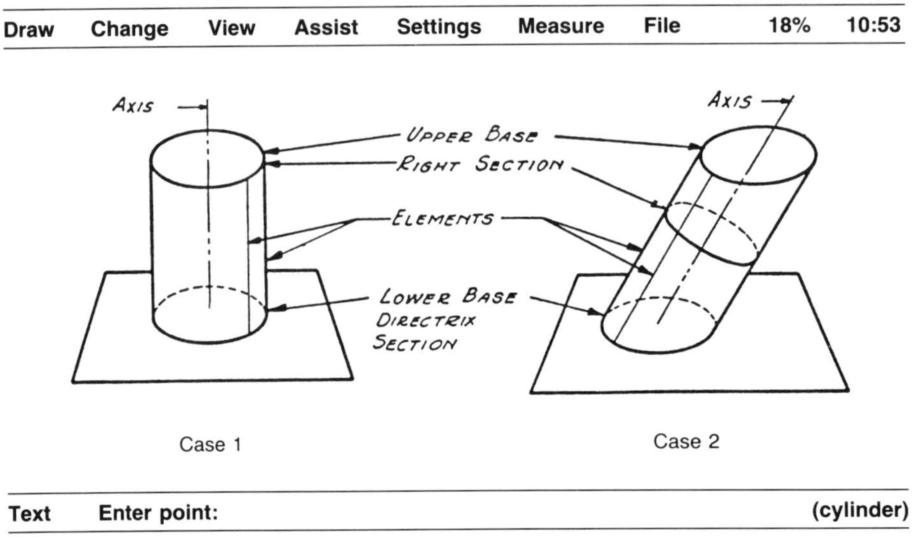

Figure 8.4 Cylinder display.

the display cylinder. A line displayed through the center of the base and parallel to the elements is called the *axis* of the cylinder. Notice in Figure 8.4 (case 1) the elements are perpendicular to the plane of the base, which is called a *right section*. In case 2 the elements are oblique to the display base.

A cylinder is displayed by the projection of a base, its axis, and the limiting elements in each projection. The shape of the base and the relative position of the axis of a cylinder are usually displayed by designating it as a right-circular cylinder or oblique-elliptical cylinder.

8.5 PROJECTING CYLINDERS

The orthographic views of a cylinder are shown in Figure 8.5. In case 1 a given curve is an oblique-elliptical cylinder, the elements of which are the horizontal and vertical projecting lines of the given curve. Therefore, any cylinders with elements perpendicular to the orthographic views are commonly referred to as *projecting cylinders*. Any and all curves lying in the surface of a projecting cylinder will have their corresponding projections in the curve of the base of the cylinders as shown in case 2. In this case a cylinder is displayed orthogonally by at least one section or base, the axis, and the limiting or contour elements in each projection.

A line *(M,O)* through any point of the known section or base and parallel to the axis is an element of the cylinder. Any point on any position of the generating line is also a point on the surface of the cylinder.

8.6 CONES

A cone is displayed by a straight line moving so that it constantly touches a given curve and passes through a fixed point not in the plane of the curve as shown in Figure 8.6. The given curve is called the *directrix*, the given point is called the

Draw	Change	View	Assist	Settings	Measure	File	31%	9:42

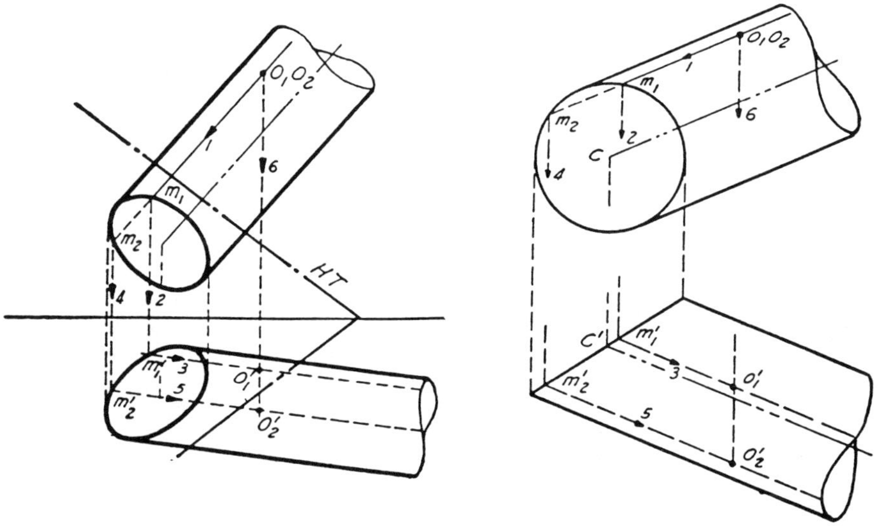

Case 1 Case 2

Text	Enter point:	(orthogrh)

Figure 8.5 Orthographic cylinders.

Draw	Change	View	Assist	Settings	Measure	File	29%	10:24

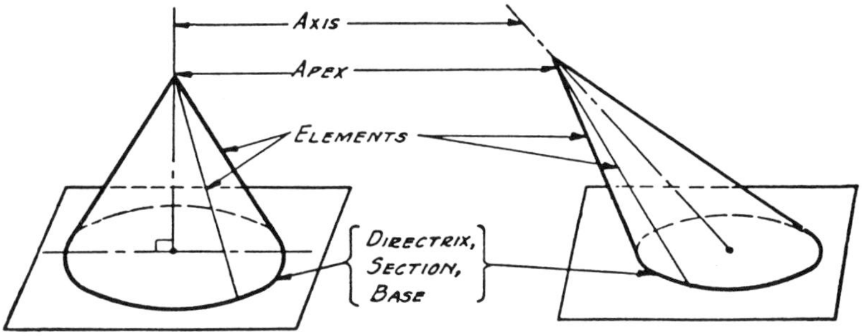

Case 1 Case 2

Text	Enter point:	(conedisp)

Figure 8.6 Cone displays.

apex, and the different positions of the generating line are called the *elements* of the cone.

If a given plane intersects all the elements of the cone, the curve of intersection is called the *base* of the cone. The straight line joining the apex with the center of the base is called the *axis* of the cone. As indicated in Figure 8.6, a cone is called a *right cone* or an *oblique cone,* depending on whether the axis is perpendicular to or oblique to the base. The shape of the base and relative position of the axis of a cone is indicated by displaying a right-circular or oblique-elliptical database.

8.7 PROJECTING CONES

A cone is displayed orthogonally by at least one section or base, the axis, and the two contour elements in each projection as shown in Figure 8.7. All positions of the generating line lie wholly on the surface of the cone. A line *(MO)* through any point of the known section, or base, and the intersection of two known elements, or apex, is an element.

8.8 GEOMETRIC CONES

A cone is displayed as a wireform in Figure 8.7 and as a geometric solid in Figure 8.8. Compare the two cones and notice that a wireform contains a number of

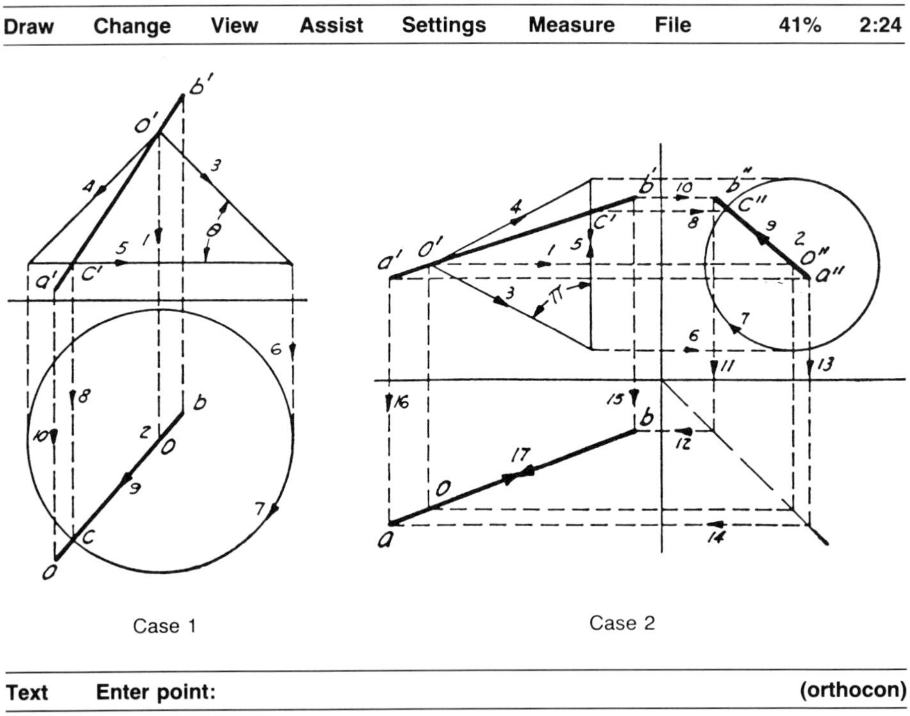

Figure 8.7 Orthographic cones.

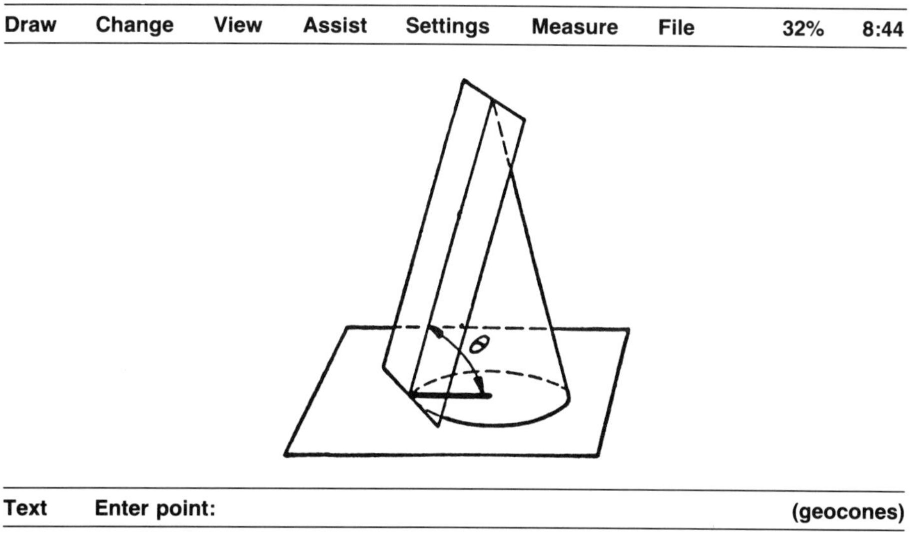

Figure 8.8 Geometric cone.

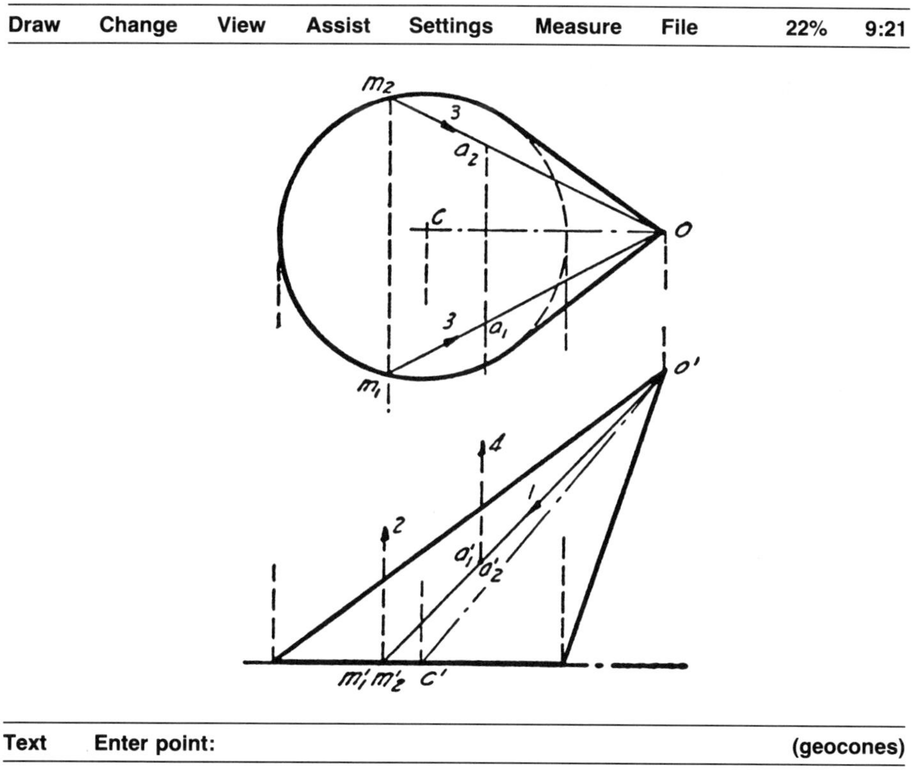

Figure 8.9 Geometric cone construction.

generating lines on the surface of the cone, while the geometric cone is identified by a tangent plane (Chapter 9). Geometric solids are constructed for display as shown in Figure 8.9.

8.9 DOUBLE-CURVED SURFACES

Thus far in this chapter we have discussed single-curved surfaces/solids. There may be an infinite number of surfaces/solids, determined by an infinite number of mathematical equation–based databanks. Those used in this chapter are the circle, ellipse, parabola, and hyperbola, called *conic sections*.

The double-curved surfaces/solids are displayed by lines that do not lie wholly in the same plane. They are three-dimensional or space shapes. A typical example is the helix shown in Figure 8.10.

8.10 HELIX

The helix is displayed by moving a curved generatrix around an axis and parallel to that axis at the same time. The generatrix usually moves at a uniform rate. If it remains a fixed distance away from the axis, a cylindrical helix is displayed. If the generatrix moves on the surface of a cone as it goes around the axis, it displays a conical helix.

Draw	Change	View	Assist	Settings	Measure	File	21%	1:13

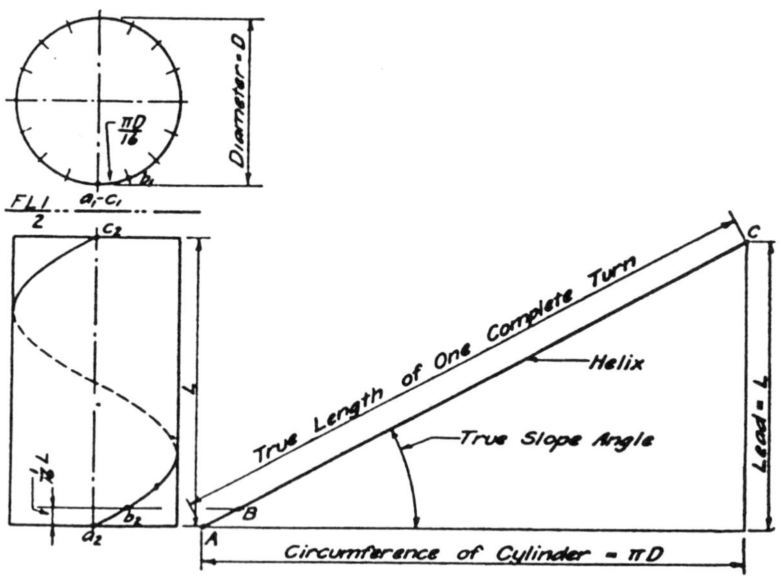

Text	Enter point:	(helixdis)

Figure 8.10 Helix displays.

Figure 8.10 shows the orthogonal views of a right-hand cylindrical helix, which is displayed by a point starting from *A* and traveling once round the cylinder whose diameter is *D* while it is traveling a distance *L* parallel to the axis of the cylinder. The distance *L* is called the *lead* of the helix. The pitch of a helix on a screw thread, for example, is the distance from a point on one thread to the corresponding point on the next thread.

The helix is displayed by dividing the distance around the cylinder into 16 equal parts. As the point *A* travels one-sixteenth of the distance around the cylinder, it travels one-sixteenth of the lead, displaying the position at *B*. By taking each position separately, the CADD construction is easily made, and the entire curve may be displayed as shown.

If a hard copy of the display is made, the triangular area of the circumference of the cylinder may be cut out. The hypotenuse of the cutout is the actual length of one complete turn of the helix. The angle shown is the true slope angle of the helix. Try wrapping the cutout to form a cylinder. The hypotenuse forms the helix.

8.11 CONVOLUTE

The convolute is shown displayed in Figure 8.11. Here a convolute was displayed by assuming it is composed of a series of triangles and by laying down the true size of these triangles in successive order. If the convolute has a constant curvature and a constant slope, the process is easier. The equation is the radius of the convolute is equal to the radius of the cylinder divided by the square of the cosine of the

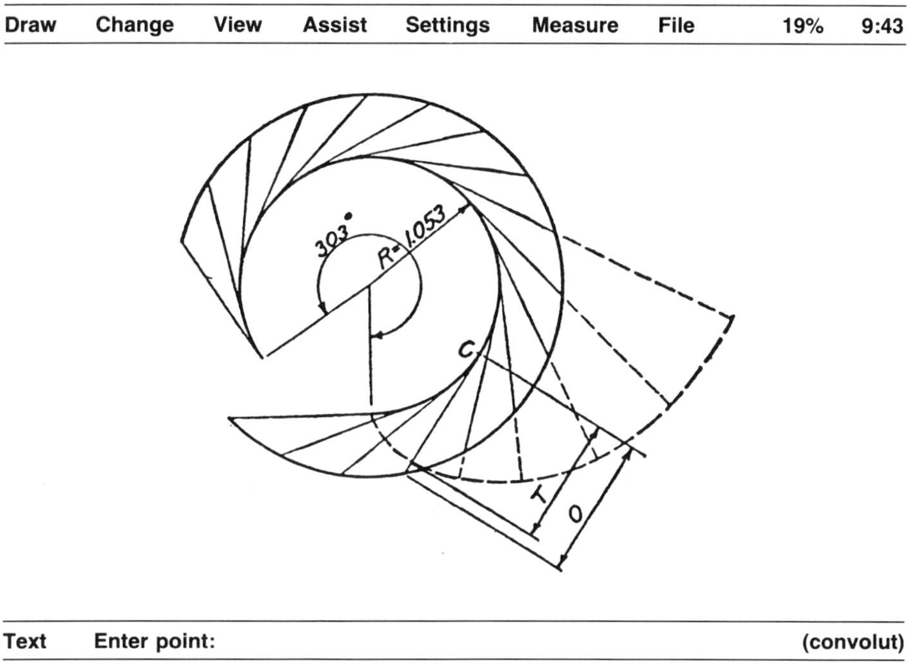

Figure 8.11 Convolute display.

slope angle. Using this, the radius is 1.053 in Figure 8.11. Sixteen divisions are used and at each division point a tangent is displayed as shown at *C*. The distance *(T* or *O)* is equal to its slope length, or true length. The ends of the tangents are then connected to form the outside edge of the convolute.

8.12 HELICAL CONVOLUTE

Figure 8.12 is the orthogonal display of a helical convolute. Here *Z* and *M* are converted to true length *(TO)*, and *NX* represents 1/16 of the helical lead of the convolute.

8.13 HELICOID

The helicoid is a warped surface that is displayed as a wireform solid in Figure 8.13. The wireform is composed of straight lines displayed about an axis in such a way that any two points on a wireform line describe concentric helices having the same lead. The generatrix need not intersect the axis, although it nearly always

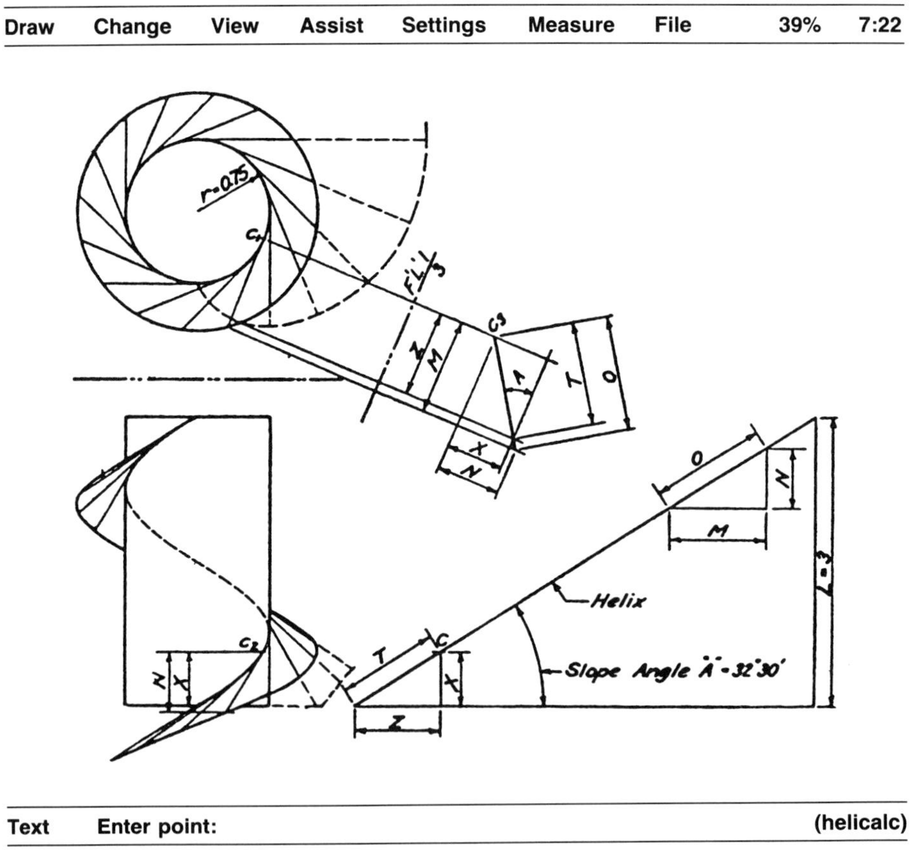

Figure 8.12 Helical convolute display.

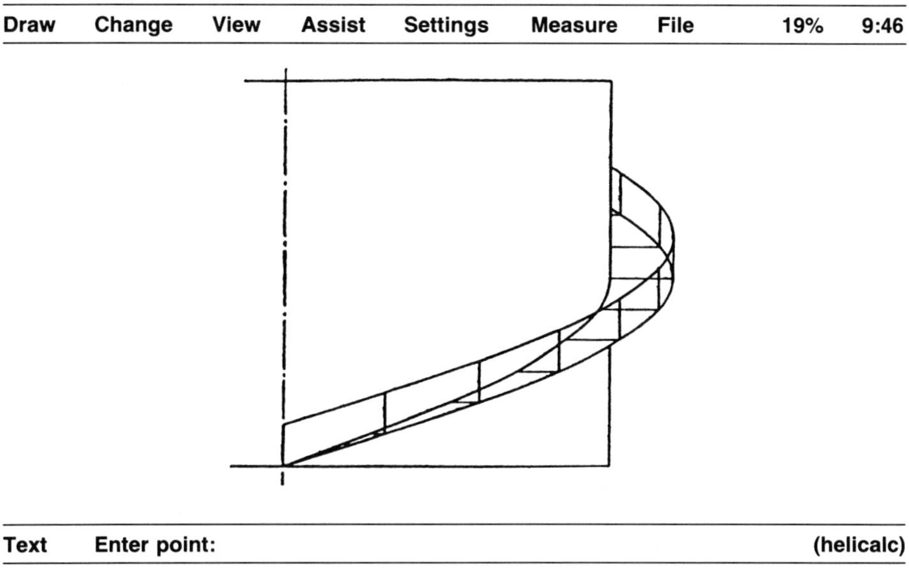

Figure 8.13 Helicoid display.

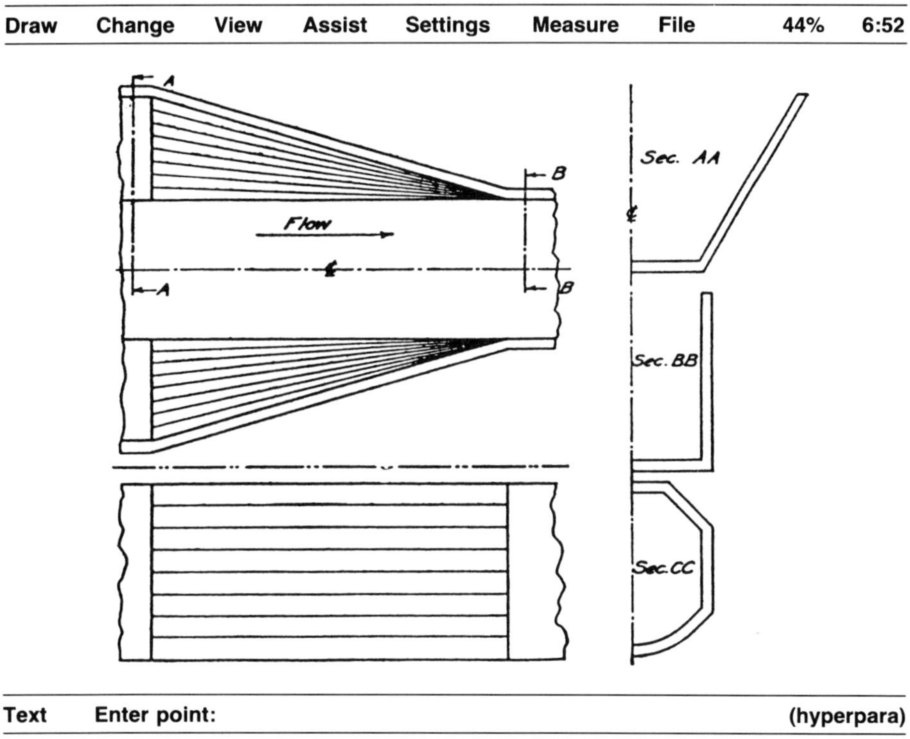

Figure 8.14 Hyperbolic paraboloid display.

does. In Figure 8.13 it intersects the axis at 90 degrees and the resulting display is called a right helicoid. Examples are the wearing surface of a square screw thread, a screw-conveyor blade, or walking ramp.

8.14 HYPERBOLIC PARABOLOID

The hyperbolic paraboloid is a warped and double-ruled surface that is displayed as a wireform solid in Figure 8.14. The wireform is composed of two straight lines displayed so that they constantly touch two nonparallel, nonintersecting lines and remain parallel to a directrix. This shape has the appearance of being twisted. In Figure 8.14, the wireform is displayed by showing the two straight-line directrices and several positions of the generatrix.

8.15 CONOID

The conoid is a warped surface that is displayed as a wireform solid in Figure 8.15. The wireform is composed of a warped surface displayed by a straight line moving so that it touches one straight-line directrix and one curved-line directrix and remains parallel to some plane directrix. The two linear directrices must lie in the same plane. The curved-line directrix may be either an open or a closed curve.

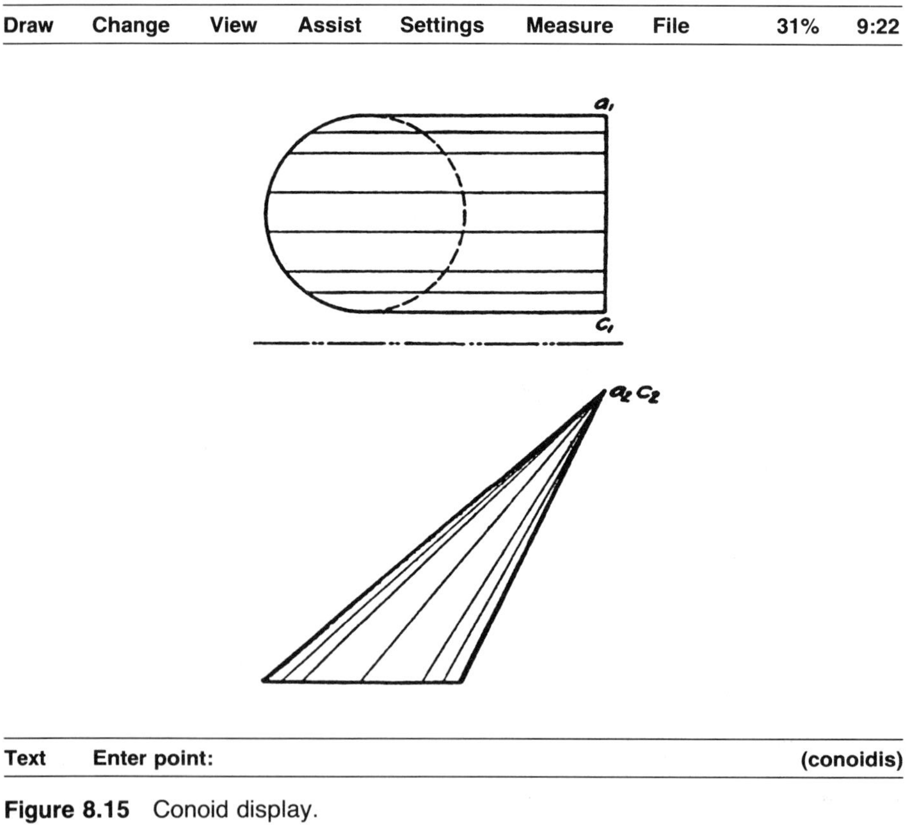

Figure 8.15 Conoid display.

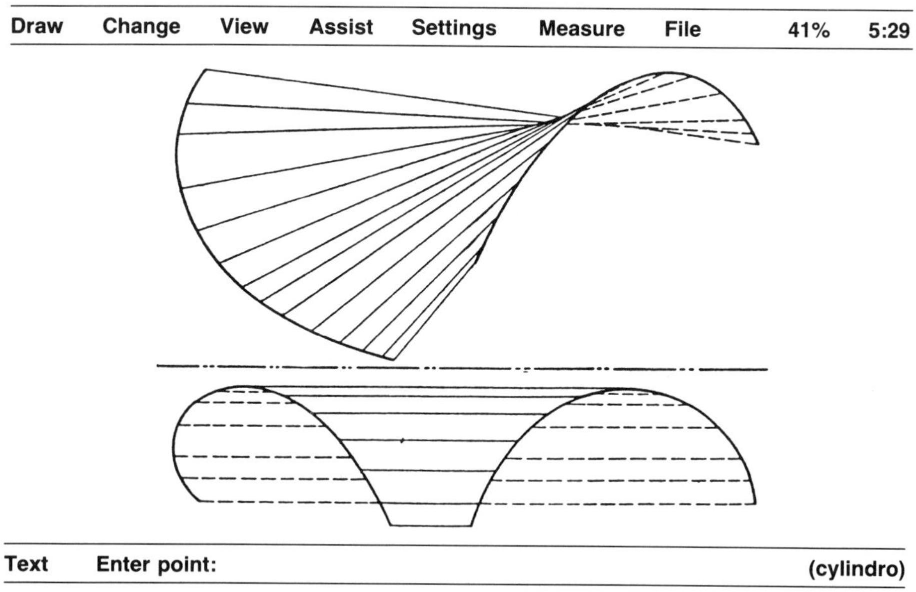

Figure 8.16 Cylindroid display.

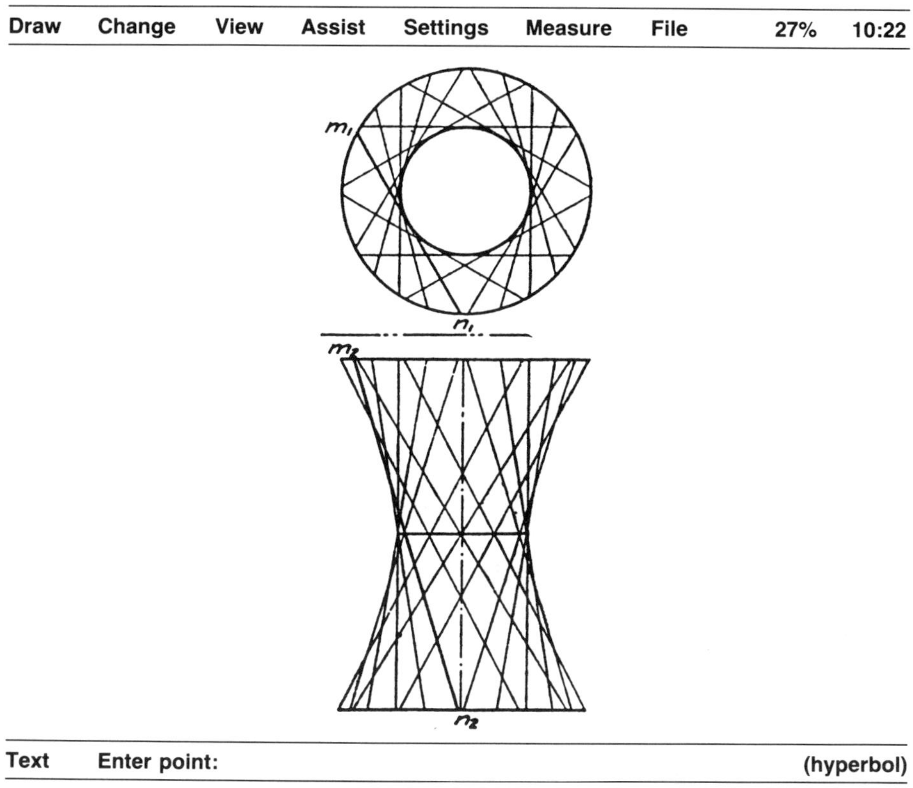

Figure 8.17 Hyperboloid display.

8.16 CYLINDROID

The cylindroid is a warped surface that is displayed as a wireform solid in Figure 8.16. The wireform is composed of a straight line moving so that it touches two curved-line directrices and remains parallel to a plane directrix. The two curved-line directrices must lie in the same plane. They may be either closed or open curves.

8.17 HYPERBOLOID

The hyperboloid is a warped surface that is displayed as a wireform solid in Figure 8.17. The wireform is composed of revolving one straight line *(MN)* about another line as an axis, providing these two lines are not parallel or intersecting. Any section of this wireform along the axis will be a hyperbola, and any plane section that is at right angles will be a circle. The smallest circular section is called the *circle of the gorge*.

8.18 SPHERE

A sphere is a double-curved surface that is displayed as a wireform solid in Figure 8.18. The wireform is composed of revolving a circle about a line passing through its center. If circles are revolved about one of its diameters as an axis, the solid generated will be the sphere shown in Figure 8.18. A sphere will appear exactly the same in all orthographic views. It always appears to be the true size of the circle by which it is displayed.

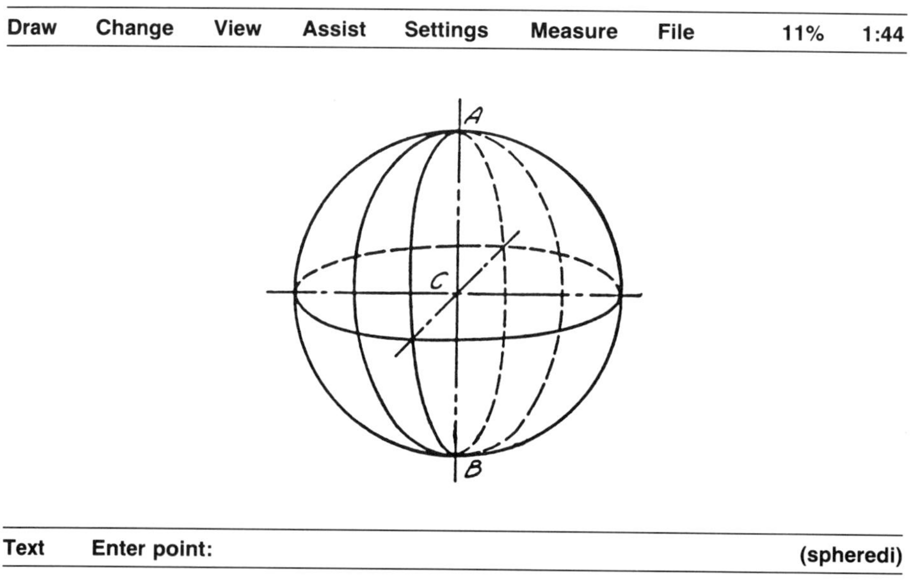

Figure 8.18 Sphere display.

Draw Change View Assist Settings Measure File 24% 4:15

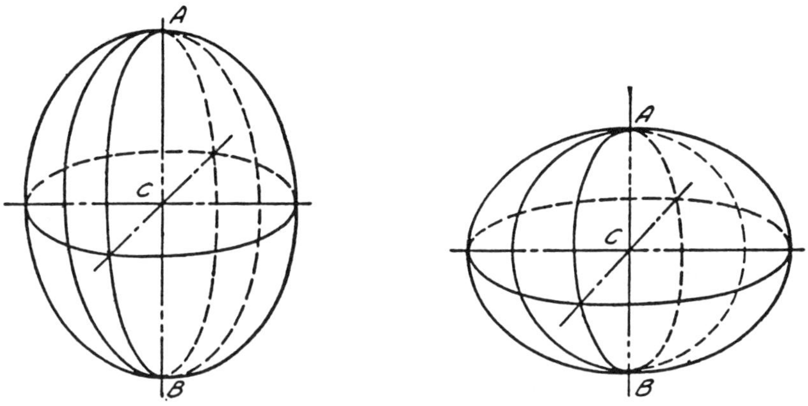

Case 1 Case 2

Text Enter point: (ellipsoi)

Figure 8.19 Ellipsoid displays.

Draw Change View Assist Settings Measure File 29% 6:25

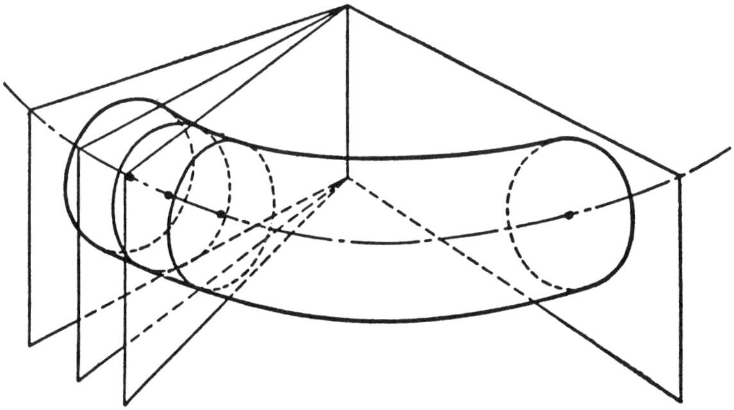

Text Enter point: (torusdis)

Figure 8.20 Torus display.

8.19 ELLIPSOID

An ellisoid is a double-curved surface that is displayed as a wireform solid in Figure 8.19. The wireform is composed of revolving an ellipse about a line passing through its center. If an ellipse is revolved about its major axis, the solid generated is called a *prolate spheroid*. If an ellipse is revolved about its minor axis, it is called an *oblate spheroid*. Case 1 is a prolate and case 2 is an oblate spheroid.

8.20 TORUS

A torus is a double-curved surface that is displayed as a wireform solid in Figure 8.20. The wireform is composed of revolving a circle about a line in the plane of the circle but not intersecting the circle. If the axis does not intersect the circle, an open torus called an *annulus* is displayed with a hole through its center. The wireform is produced by the rotation of one circle about a curved axis.

8.21 CHAPTER PROBLEMS

The problems appearing at the end of this chapter have been selected to introduce the basic types of broad, general problems encountered by an engineer or engineering technician using a CADD display system. These problems emphasize the interdisciplinary aspects of constructive geometry, which require a number of related pieces of CADD software. The software used for the production of problems in this chapter can be ordered from Autodesk, Inc.

PROBLEM SET 1 *Straight line plane surfaces*

In each problem display the plane defined.

1.1 Plane *A* (Fig. 8.1 [1])
1.2 Plane *B* (Fig. 8.1 [2])
1.3 Plane *T* (Fig. 8.2 [1])
1.4 Plane *T* (Fig. 8.2 [2])
1.5 Plane *T* (Fig. 8.3 [1])
1.6 Plane *B* (Fig. 8.3 [2])
1.7 Plane *C* (Fig. 8.3 [3])
1.8 Plane *A* (Fig. 8.4)
1.9 Plane *HT* (Fig. 8.5 [1])
1.10 Plane *A* (Fig. 8.6 [1])

PROBLEM SET 2 *Cylinder displays*

In each problem display the cylinder defined.

2.1 *T* (Fig. 8.2 [2])
2.2 *B* (Fig. 8.3 [2])
2.3 *A* (Fig. 8.4 [1])
2.4 *B* (Fig. 8.4 [2])
2.5 *P* (Fig. 8.5 [1])
2.6 *C* (Fig. 8.5 [2])

2.7 *A* (Fig. 8.10)
2.8 *C* (Fig. 8.12)
2.9 *H* (Fig. 8.13)
2.10 *X* (Fig. 8.15)

PROBLEM SET 3 *Cone displays*

In each problem display the cone defined.

3.1 *A* (Fig. 8.3 [3])
3.2 *B* (Fig. 8.3 [3])
3.3 *A* (Fig. 8.6 [1])
3.5 *B* (Fig. 8.6 [2])
3.6 *O* (Fig. 8.7 [1])
3.7 *O* (Fig. 8.7 [2])
3.8 *O* (Fig. 8.8)
3.9 *C* (Fig. 8.9)
3.10 *A* (Fig. 8.15)

PROBLEM SET 4 *Helix/convolute displays*

In each problem display the object defined.

4.1 Helix (Fig. 8.10)
4.2 Convolute (Fig. 8.11)
4.3 Helical convolute (Fig. 8.12)
4.4 Helicoid (Fig. 8.13)
4.5 Hyperbolic paraboloid (Fig. 8.14)
4.6 Conoid (Fig. 8.15)

PROBLEM SET 5 *Double-curved object displays*

In each problem display the object defined.

5.1 Cylindroid (Fig. 8.16)
5.2 Hyperboloid (Fig. 8.17)
5.3 Sphere (Fig. 8.18)
5.4 Ellipsoid (Fig. 8.19)
5.5 Torus (Fig. 8.20)

9 Tangent Planes and Solids

This chapter continues the CAD/CAE relationship between the plane and the solid. Prior to this chapter the relationship has been a separate plane or a separate solid. The series of displays shown in this chapter consists of the linking of planes and solids to form useful CAD/CAE objects. Each of the display situations is a section heading; under each heading is an example of the CADD construction required for the display.

If planes and solids are joined as objects that touch at a single point along a line or surface as defined in Chapter 4, then this chapter would fall short of its desired application. As pointed out in Chapter 8, however, a plane is generated by a line moving in a straight or constantly changing direction. Therefore, solids can be generated by a straight or curved plane moving in a direction.

9.1 PLANES TANGENT TO A SOLID

Planes are tangent to a solid along a straight line as shown in Figure 9.1, case 1. This is usually the most common case used in CAE applications. A plane can be tangent to a solid at a point as shown in case 2, however. The tangent plane (*T*) in Figure 9.1 is composed of a series of straight lines.

9.2 SINGLE-CURVED SURFACES

Single-curved surfaces are best displayed as a cylinder or cone. A plane surface tangent to either of these is defined by any two lines in it. Therefore, a plane that is tangent to the surface of a cone or cylinder contains one of the elements of the surface. This line is one of the two required to define the plane. Therefore a line tangent to the surface at any point of the line of tangency is the second line required to define the plane.

9.3 TANGENT PLANES THROUGH A POINT

A common CAD/CAE construction is the display of a plane through a given point on a given cone or cylinder, tangent to the surface. A plane is shown in Figure 9.2 and is defined as *XPA*. This plane is tangent to the surface of the cone through point *O* and contains one of the elements of the surface of the cone. This line is also one of the two required to define the plane surface. The required plane is defined by a second line *PA*.

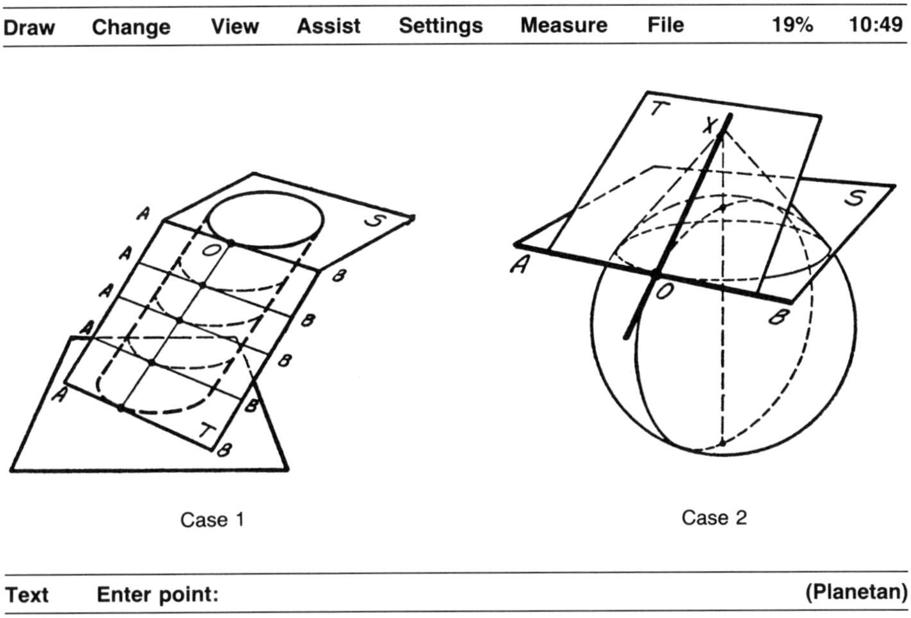

Figure 9.1 Plane tangent to solids.

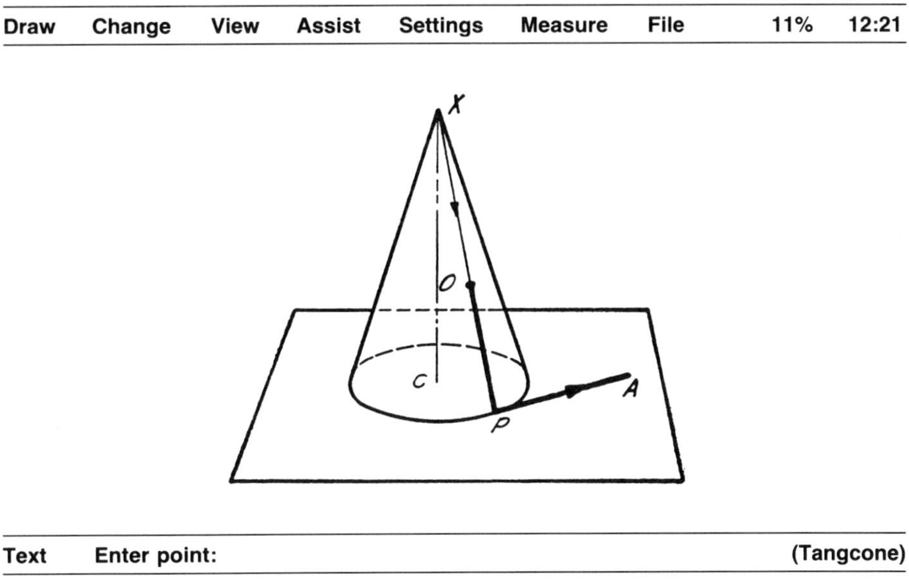

Figure 9.2 Plane tangent to a cone.

9.4 DISPLAY CONSTRUCTION FOR CONES

In Figure 9.3, a cone is displayed as *XC* in two orthographic views. In the frontal view the base of the cone appears as a straight line. Two planes are displayed tangent to the cone at points *O*(1) and *O*(2). The steps of construction are numbered in Figure 9.3 to illustrate the projections of the elements *XO*(1) and *XO*(2) through *O*′(1)*O*′(2) and *X*′ extended until they intersect the base at *P*′(1) and *P*′(2).

P′(1) and *P*′(2) are the projections of the two points in which the elements pierce the plane of the base of the cone. Their *H* projections are at *P*(1) and *P*(2), and the *H* projections of the elements are *P*(1)*X* and *P*(2)*X*.

The lines tangent to the base at *P*(1) and *P*(2) are *P*(1)*A* and *P*(2)*B*. The required planes *R* and *S* are then constructed by lines *XP*(1), *P*(1)*A* and *XP*(2), *P*(2)*B*. The trace of plane *R* (step 6) is shown, and the trace of plane *S* (step 8) is displayed.

9.5 TANGENT PLANE/EXTERNAL POINT

Another common CAD/CAE construction is the display of a plane through a given point external to the cone or cylinder, tangent to the surface. A plane is shown in Figure 9.4 and is defined as *XAC*. The plane *CPX* is tangent to the surface of the

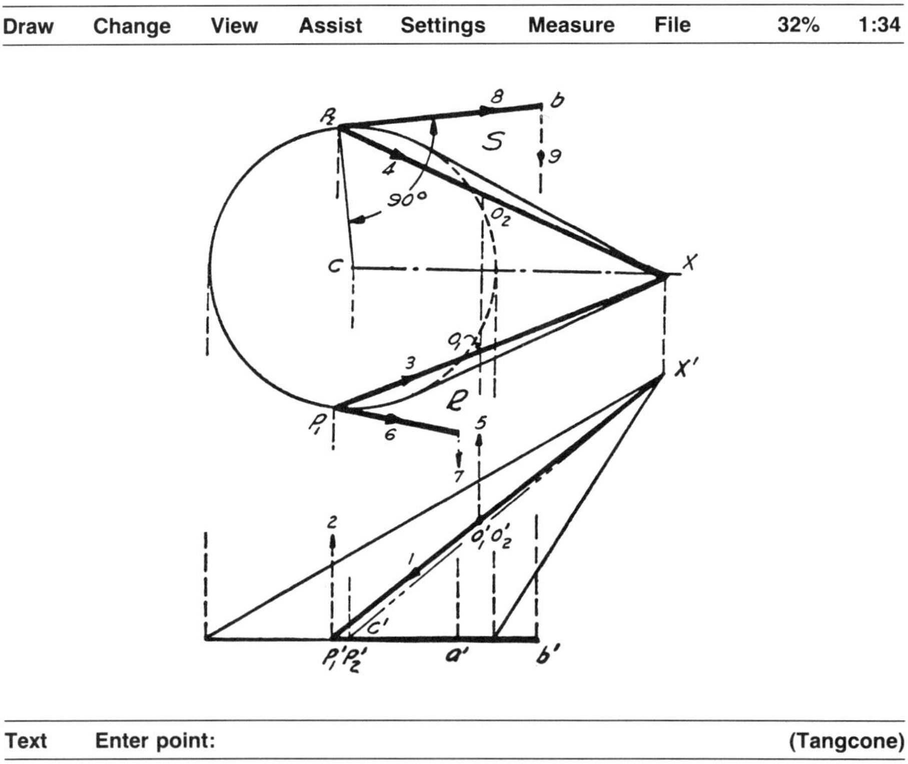

Figure 9.3 Display construction.

Draw Change View Assist Settings Measure File 12% 2:12

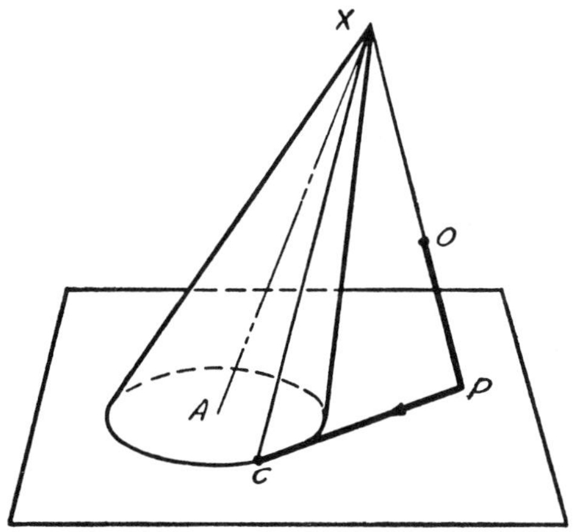

Text Enter point: (Tangcon2)

Figure 9.4 Tangent planes through external points.

Draw Change View Assist Settings Measure File 28% 1:43

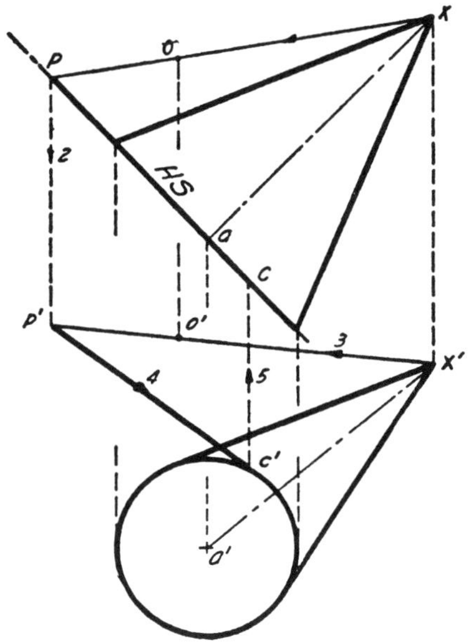

Text Enter point: (Tangcon3)

Figure 9.5 Steps of construction.

cone through point *O* and contains one of the elements of the surface of the cone (*XC*). This line is not one of the two required to define the plane surface, however. The required plane is defined by a second line *PC* and is the trace of the plane surface. See Figure 9.5 for the steps of construction.

9.6 TANGENT PLANE/EXTERNAL LINE

A useful CAD/CAE construction is the display of a plane through a given line external to the cone or cylinder, tangent to the surface and parallel to the line. A shaded plane is shown in Figure 9.6 and is defined as *XCP*. This plane *XCP* is tangent to the surface of the cone and parallel to line *AB*. The line *AB* is not one of the two required to define the plane surface, however. The required plane is defined by a second line *PC* and is the trace of the plane surface. See Figure 9.7 for the steps of construction.

9.7 PLANE/LINE AND CYLINDER

Another useful CAD/CAE construction is the display of a plane parallel to a given line external to the cylinder, tangent to the surface of the cylinder. A plane is displayed in Figure 9.8 and is defined as *FDE*. This plane *FDE* is tangent to the surface of the cylinder and parallel to line *AC*. The line *AC* is not one of the two required to define the plane surface, however. The required plane is defined by a second line *DE*. A second plane can be constructed parallel to the tangent plane by the insertion of points *P*(1) and *P*(2). See Figure 9.9 for the steps of construction.

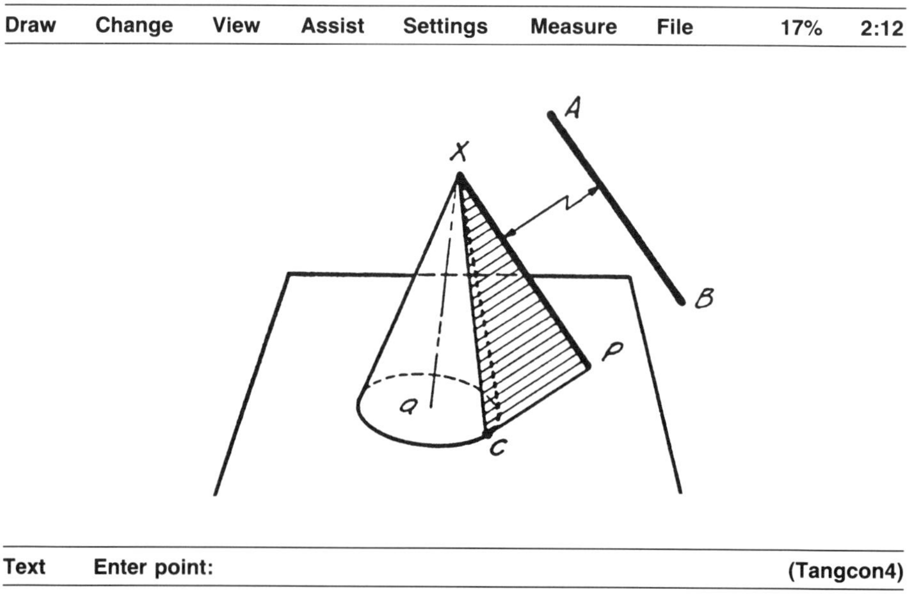

Figure 9.6 Tangent plane and external line.

Draw **Change** **View** **Assist** **Settings** **Measure** **File** **31%** **3:27**

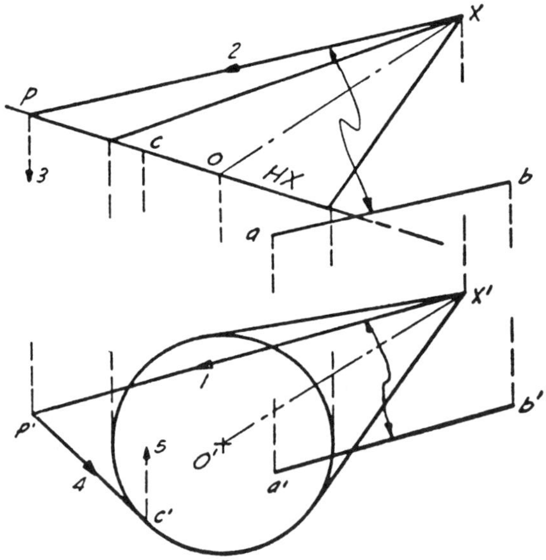

Text **Enter point:** **(Tangcon5)**

Figure 9.7 Steps of construction.

Draw **Change** **View** **Assist** **Settings** **Measure** **File** **14%** **3:22**

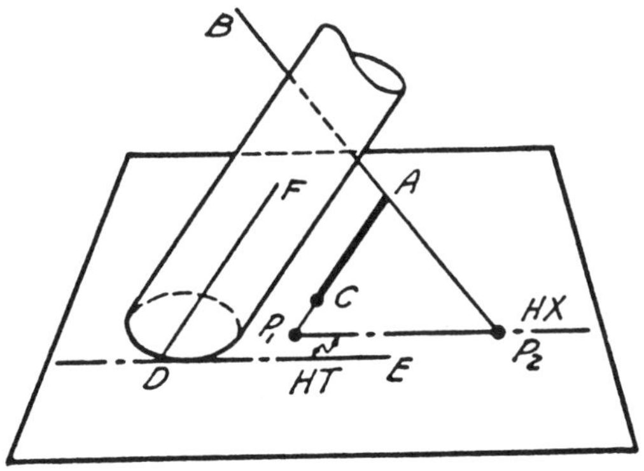

Text **Enter point:** **(Tangcycl)**

Figure 9.8 Tangent plane and external plane.

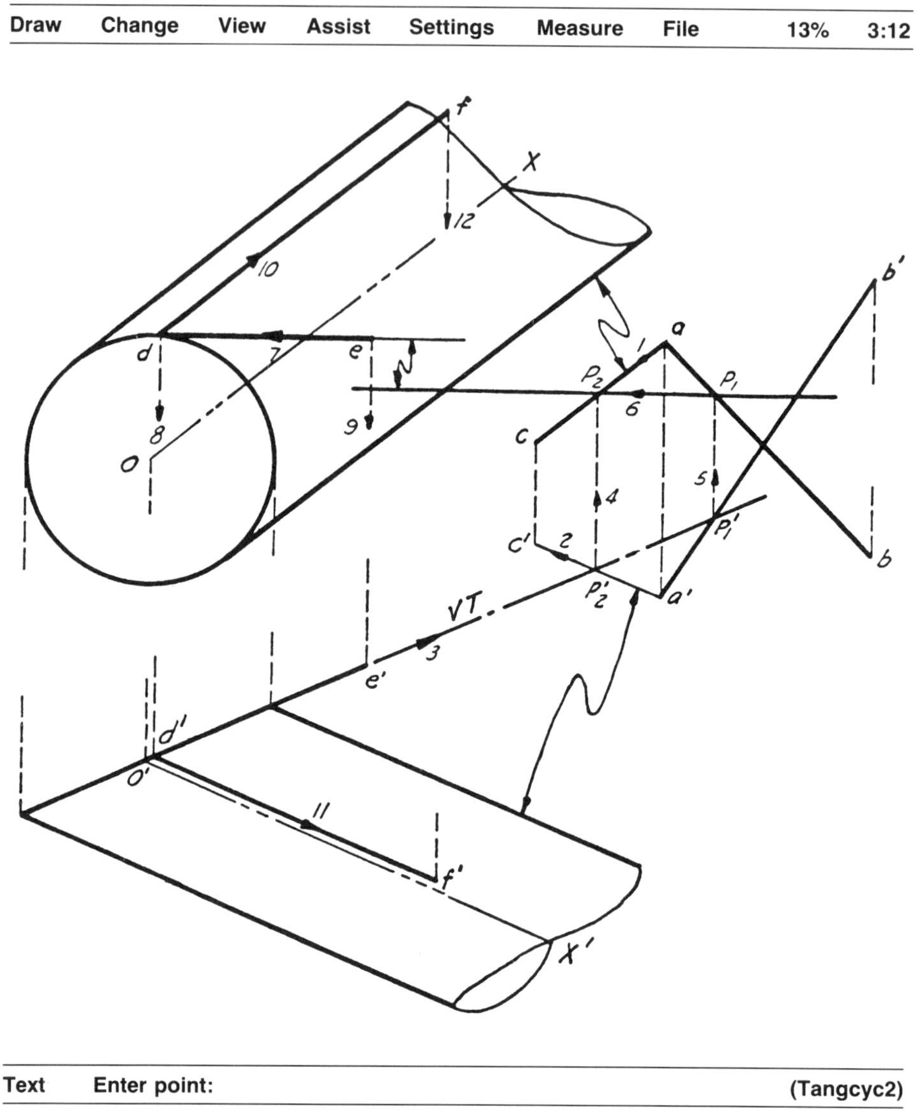

Figure 9.9 Steps of construction.

9.8 PLANE/LINE AND ANGLE

Another useful CAD/CAE construction is the display of a plane through a given line that makes a specified angle with a given plane. A shaded plane is displayed in Figure 9.10 and is defined as *OHA*. This plane *OHA* is tangent to the surface of the cone at line *OH*(2). The line *AB* is one of the lines required to define the plane surface and is at the specified angle. The tangent plane is defined by a second line *H*(1),*H*(2). See Figure 9.11 for the steps of construction.

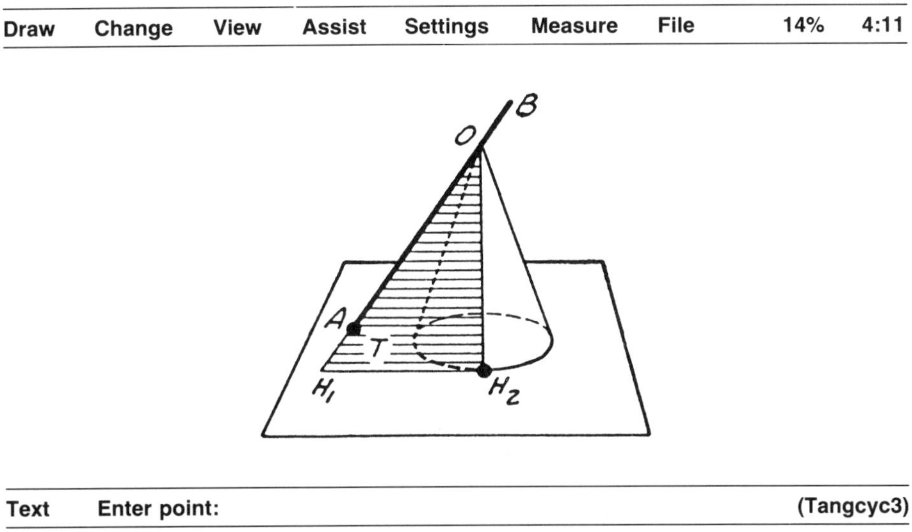

Figure 9.10 Tangent plane and external angle.

9.9 DOUBLE-CURVED SURFACES

Double-curved surfaces are best displayed as a spheroid. A plane surface tangent to a sphere is defined by a point and a line on it. Therefore a plane that is tangent to the surface of a sphere contains one of the elements of the surface. This line is one of the two required to define the plane. Therefore a line tangent to the surface at any point of the line of tangency is the second line required to define the plane as shown in Figure 9.12.

9.10 TANGENT PLANE THROUGH A POINT

A common CAD/CAE construction is the display of a plane through a given point on a given sphere, tangent to the surface. A shaded plane is shown in Figure 9.12 and is defined as *APX*. This plane is tangent to the surface of the sphere through point *P* and contains one of the elements of the surface of the sphere. This point is also one of the two required to define the plane surface. The required plane is defined by the second line *PA*. See Figure 9.13 for the construction.

9.11 TANGENT PLANE/EXTERNAL POINT

Another common CAD/CAE construction is the display of a plane through a given point external to the sphere, tangent to the surface. A plane is shown in Figure 9.14 and is defined as *APX*. The plane *APX* is tangent to the surface of the sphere through point *P* and contains one of the two intersecting elements of the surface of the sphere. This point is one of the two required to define the plane surface location. The required plane is defined by a second line *PA* and is the trace of the plane surface. See Figure 9.15 for the steps of construction.

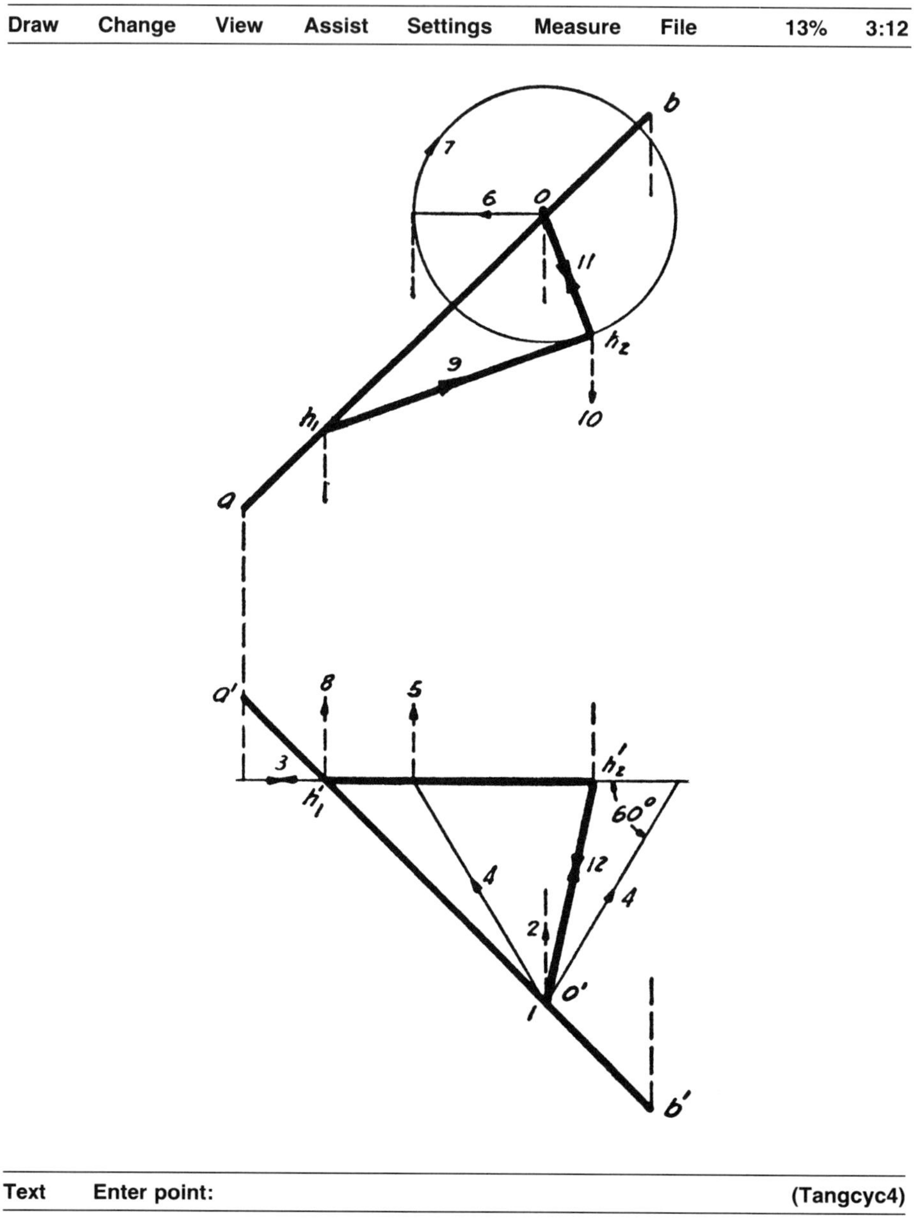

Figure 9.11 Steps of construction.

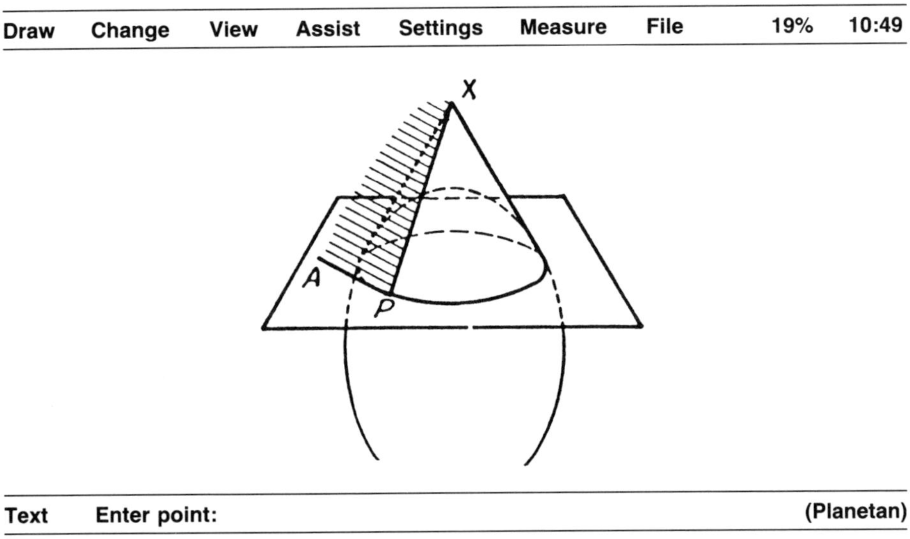

Figure 9.12 Plane tangent to sphere.

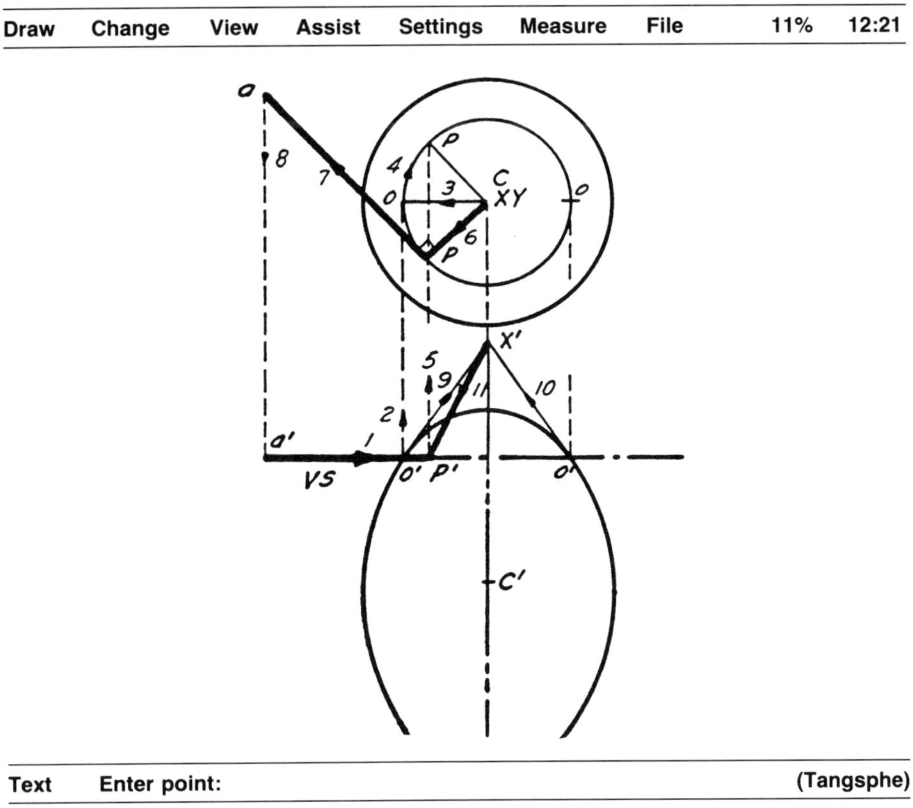

Figure 9.13 Construction of tangent to a sphere.

Draw	Change	View	Assist	Settings	Measure	File	28%	1:43

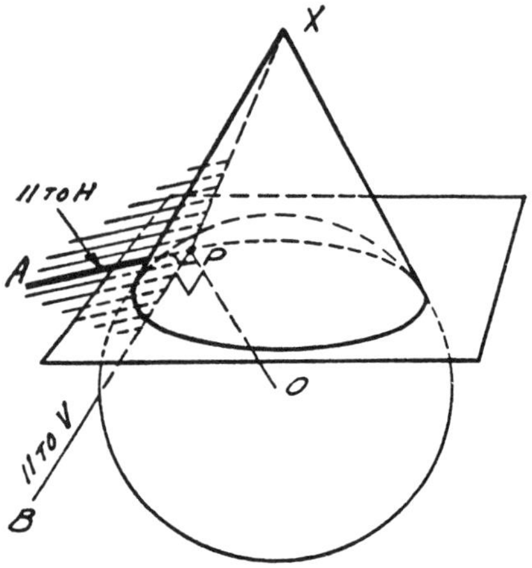

Text	Enter point:	(Tangsphe)

Figure 9.14 Tangent through points.

Draw	Change	View	Assist	Settings	Measure	File	13%	3:12

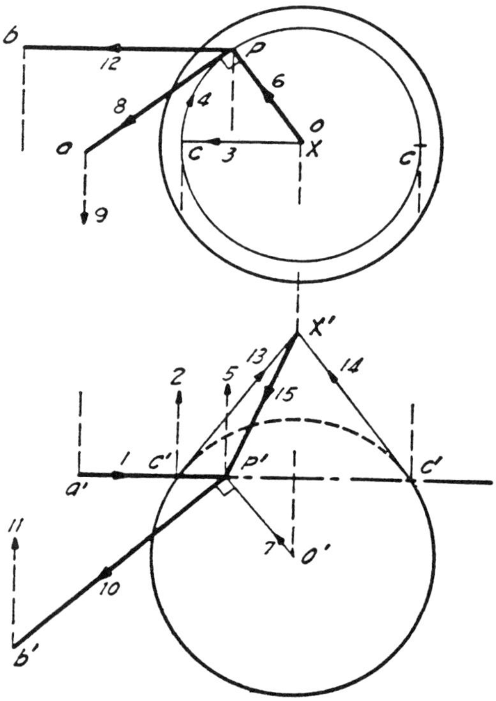

Text	Enter point:	(Tangsph2)

Figure 9.15 Steps of construction.

9.12 CHAPTER PROBLEMS

The problems appearing at the end of this chapter have been selected to introduce the basic types of broad, general problems encountered by an engineer or engineering technician using a CADD display system. These problems emphasize the interdisciplinary aspects of constructive geometry, which require a number of related pieces of CADD software. The software used for the production of problems in this chapter can be ordered from Autodesk, Inc.

PROBLEM SET 1 *Plane tangent to a solid*

In each problem display the plane tangent to the solid defined.

1.1 Plane *S* tangent to cylinder along line *O* (Fig. 9.1 [1])
1.2 Plane *T* tangent to sphere at point *O* (Fig. 9.1 [2])
1.3 Plane *PA* tangent to cone at *O* (Fig. 9.2)
1.4 Plane *R* tangent to cone at *O*(1), (Fig. 9.3)
1.5 Plane *S* tangent to cone at *O*(2), (Fig. 9.3)
1.6 Plane *CP* tangent to cone at *XC* (Fig. 9.4)
1.7 Plane *HS* tangent at base of cone (Fig. 9.5)
1.8 Plane *CP* tangent to cone (Fig. 9.6)
1.9 Plane *HX* tangent at base of cone (Fig. 9.7)
1.10 Plane *HT* tangent at base of cylinder (Fig. 9.8)

PROBLEM SET 2 *Cylinder/plane displays*

In each problem display the plane and cylinder defined.

2.1 *T* tangent to cylinder along *O* (Fig. 9.1 [1])
2.2 *Z* tangent to cylinder along *CF* (Fig. 9.8)
2.3 *X* parallel to *FD* (Fig. 9.8)
2.4 *CA* parallel to axis (Fig. 9.8)
2.5 *P* tangent to any element (Fig. 9.9)
2.6 *D* parallel to axis (Fig. 9.9)
2.7 *Q* parallel to *D* and tangent to any element (Fig. 9.9)
2.8 *T* tangent to base (Fig. 9.9)
2.9 *O* parallel to base 2 (Fig. 9.9)
2.10 *X* tangent to any element (Fig. 9.9)

PROBLEM SET 3 *Cone/plane displays*

In each problem display the plane and cone defined.

3.1 *T* tangent through point *O* (Fig. 9.2)
3.2 *S* tangent to any element (Fig. 9.3)
3.3 *R* tangent to any element (Fig. 9.3)
3.5 *P* tangent through point *O* (Fig. 9.4)
3.6 *S* parallel to the base (Fig. 9.5)
3.7 *P* tangent to any element and parallel to *AB* (Fig. 9.6)
3.8 *AB* at any angle tangent at point *O* (Fig. 9.10)
3.9 *T* tangent at 45 degrees to *O* (Fig. 9.10)
3.10 *P* tangent at 60 degrees to *O* (Fig. 9.11)

PROBLEM SET 4 *Sphere/plane displays*

In each problem display the sphere and plane defined.

4.1 *T* tangent at point *O* (Fig. 9.1 [2])
4.2 *S* perpendicular to axis (Fig. 9.1 [2])
4.3 *X* tangent to any point (Fig. 9.12)
4.4 *P* parallel through *A* (Fig. 9.12)
4.5 *S* tangent at any element (Fig. 9.13)
4.6 *V* perpendicular to axis (Fig. 9.13)
4.7 *V* parallel to axis (Fig. 9.14)
4.8 *A* parallel to *H* (Fig. 9.14)
4.9 *B* tangent through *P* (Fig. 9.14)
4.10 *AP* parallel to any element (Fig. 9.14)

PROBLEM SET 5 *Plane through point*

In each problem display the point through the defined point.

5.1 *T* through *O* (Fig. 9.1 [1] or [2])
5.2 *P* through *O* (Fig. 9.2)
5.3 *X* through *O* (Fig. 9.4)
5.4 *A* through *C* (Fig. 9.6)
5.5 *A* through *P*(1) (Fig. 9.8)
5.6 *T* through *O* (Fig. 9.10)
5.7 *A* through *P* (Fig. 9.12)
5.8 *A* through *P* (Fig. 9.14)
5.9 *A* through *X* (Fig. 9.14)
5.10 *B* through *P* (Fig. 9.15)

PROBLEM SET 6 *Plane through external point/cone*

In each problem display the cone and plane through the external point.

6.1 *P* through *A* (Fig. 9.2)
6.2 *S* through any point (Fig. 9.3)
6.3 *R* through any point (Fig. 9.3)
6.4 *X* through *O* (Fig. 9.4)
6.5 *S* through *P* (Fig. 9.5)
6.6 *P* through *B* (Fig. 9.6)
6.7 *T* through any point (Fig. 9.10)
6.8 *B* through *H* (Fig. 9.10)
6.9 *A* through any point (Fig. 9.11)
6.10 *B* through any point (Fig. 9.11)

PROBLEM SET 7 *Plane through external point/cylinder*

In each problem display the cylinder and plane through the external point.

7.1 *S* through any point (Fig. 9.1[1])
7.2 *T* through *P* (Fig. 9.1[1])
7.3 *T* through *P*(2) (Fig. 9.8)
7.4 *X* through *B* (Fig. 9.8)

7.5 *T* through any point (Fig. 9.9)
7.6 *F* through *P*(2) (Fig. 9.9)
7.7 *A* through *O* (Fig. 9.1[1])
7.8 *T* through *D* (Fig. 9.8)
7.9 *B* through any point parallel to *AC* (Fig. 9.8)
7.10 *O* through any point perpendicular to *X* (Fig. 9.9)

PROBLEM SET 8 *Plane parallel to cone*

In each problem display the cone and plane defined.

8.1 *A* to *XP* parallel to any element (Fig. 9.2)
8.2 *S* to *XP* parallel to *O*(2) (Fig. 9.3)
8.3 *R* to *XP* parallel to any element (Fig. 9.3)
8.4 *P* to *XA* parallel to *XC* through *P* (Fig. 9.4)
8.5 *Q* to *HS* parallel at any distance (Fig. 9.5)
8.6 *P* to *AB* parallel to any element (Fig. 9.6)
8.7 *H* to *XO* parallel *HX* (Fig. 9.7)
8.8 *T* to *AB* parallel at any distance (Fig. 9.10)
8.9 *Q* to *AB* parallel to any element (Fig. 9.11)
8.10 *D* to *OH* parallel at any distance (Fig. 9.11)

PROBLEM SET 9 *Plane parallel to cylinder*

In each problem display the cylinder and plane defined.

9.1 *S* to top parallel at any distance (Fig. 9.1 [1])
9.2 *T* to *O* parallel to *AB* (Fig. 9.1 [1])
9.3 *S* to base parallel to any element (Fig. 9.1 [1])
9.4 *X* to *AC* parallel at any distance (Fig. 9.8)
9.5 *T* to base parallel at any distance (Fig. 9.8)
9.6 *Q* to *XO* parallel at any distance (Fig. 9.9)
9.7 *E* to *DF* parallel to any element (Fig. 9.9)
9.8 *F* to *DE* parallel to any distance (Fig. 9.9)
9.9 *O* to base parallel to any element (Fig. 9.9)
9.10 *C* to *AB* parallel at any distance (Fig. 9.9)

PROBLEM SET 10 *Plane parallel to sphere*

In each problem display the sphere and plane defined.

10.1 *T* parallel at *O* (Fig. 9.1 [2])
10.2 *S* parallel at *AB* (Fig. 9.1 [2])
10.3 *A* parallel at *XP* (Fig. 9.12)
10.4 *P* parallel at *AX* (Fig. 9.12)
10.5 *P* parallel at *C* (Fig. 9.13)
10.6 *C* parallel at *VS* (Fig. 9.13)
10.7 *P* parallel at *O* (Fig. 9.14)
10.8 *A* parallel at *P* (Fig. 9.14)
10.9 *B* parallel at *P* (Fig. 9.15)
10.10 *C* parallel at *O* (Fig. 9.15)

10 Intersection of Planes and Solids

This chapter introduces the CADD relationship between the intersection of a plane and a solid. Prior to this chapter the relationship has been the line of intersection of any two planes. The series of display solids shown in this chapter consists of the linking of planes to form solids. Each of the display situations is a section heading; under each heading is the CADD construction required for the display.

The surface of intersection of a plane and solid is the locus of all points common to the matching surfaces. If a line on one matching surface intersects a line on another surface, their intersection is a point common to the intersection of the plane with the solid. If two intersecting solids are cut by a plane, the lines cut from them may intersect, since they lie in a common plane.

10.1 CUTTING PLANE METHOD

A simple and generally used method of determining the intersection of solids involves the passing of cutting planes in such a manner as to intersect the surfaces in lines that project on the coordinate planes either as straight lines or circles. In applying this method the following facts of solid geometry should be remembered:

1. Cutting planes passing through the apex of a cone intersect the surface in elements as in Figure 10.1, case 1.
2. Cutting planes parallel to the axis of a cylinder intersect the surface in elements as in Figure 10.1, case 2.
3. Cutting planes parallel to the base of a cone or cylinder intersect the surface in curves similar to or equal to the base as in Figure 10.2, case 1.
4. Cutting planes perpendicular to the axis of a solid generated by revolution intersect in one or more circles as in Figure 10.2, case 2.
5. Planes through the axis of a revolved solid form meridian lines as in Figure 10.2, case 3.

10.2 VISIBILITY

The various surfaces and solids treated in CAD/CAE geometric modeling are considered opaque. In describing solids on the coordinate planes there is thus a need for conventional methods to designate what is on the side toward the observer, and hence visible, and what is on the side away from the observer, and hence invisible. Points on the surfaces of the various solids may be visible in some views and

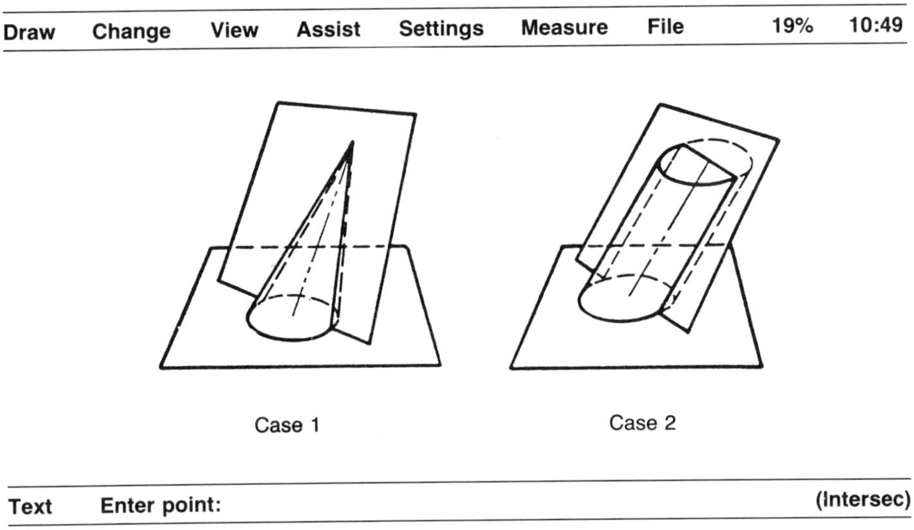

Figure 10.1 Intersection of a cone or cylinder.

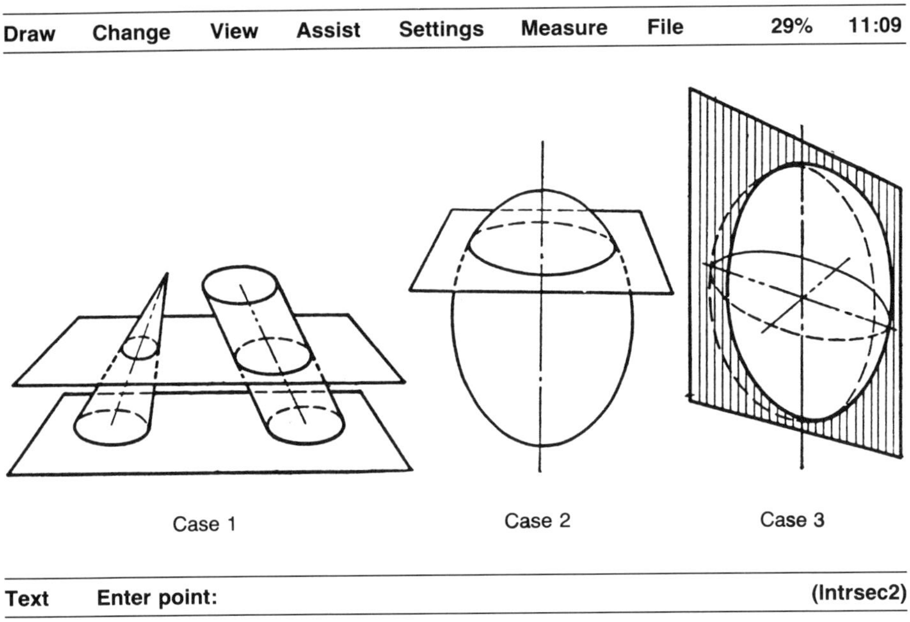

Figure 10.2 Intersection of solids.

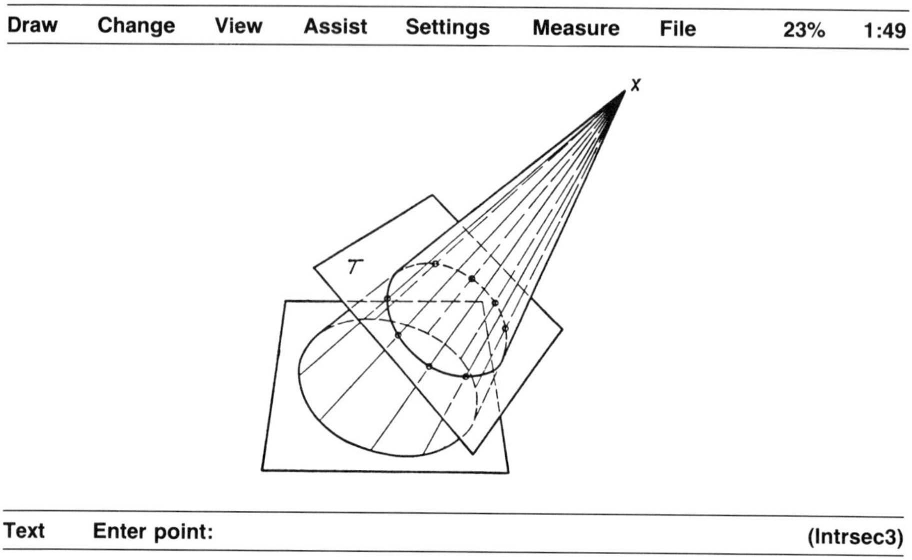

Figure 10.3 Intersection of a cone and plane.

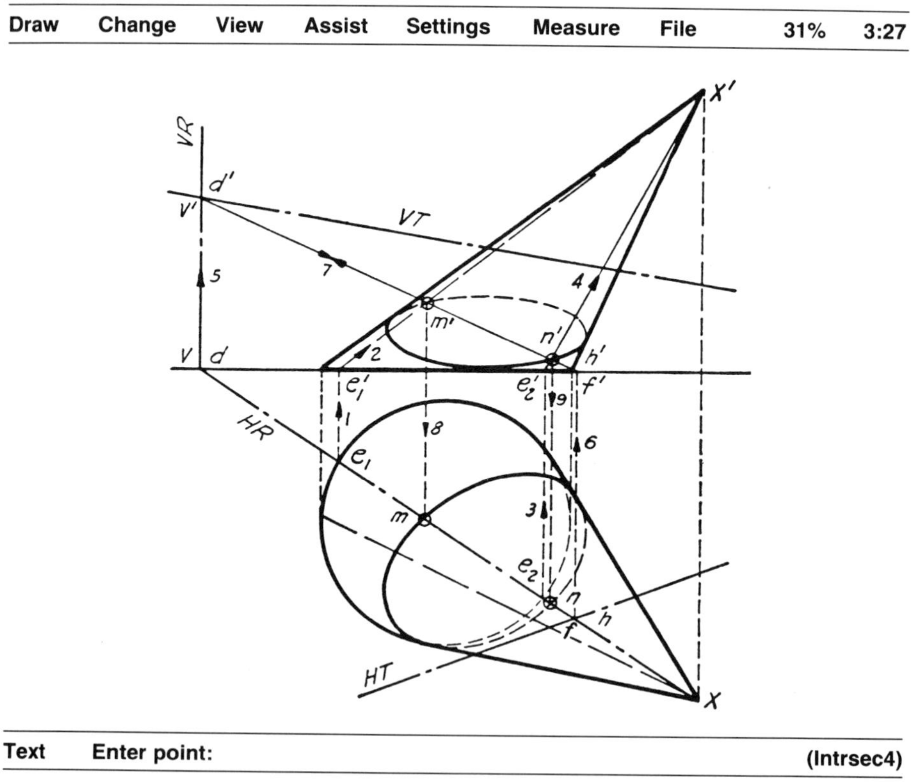

Figure 10.4 Steps of construction.

invisible in others. There arises therefore the need for a method of determining if a point is on the side toward or away from the observer.

10.3 PLANE AND CONE

The intersection of a plane and a cone is displayed in Figure 10.3 by a series of points common to the plane and the surface of the cone. Points are displayed as the intersections of lines, which in turn display planes.

Points are displayed at the desired intervals to display the required line of intersection by passing a series of plane surfaces such that they intersect the plane and cone in straight lines. The intersections of each set of lines will be points of the intersection. See Figure 10.4 for construction.

10.4 PLANE OF INTERSECTION: CONE

The plane of intersection is displayed in Figure 10.5 by a series of points common to the plane (*T*) and the elements of the cone. Points *MN* are displayed as the intersections of lines, which in turn display planes.

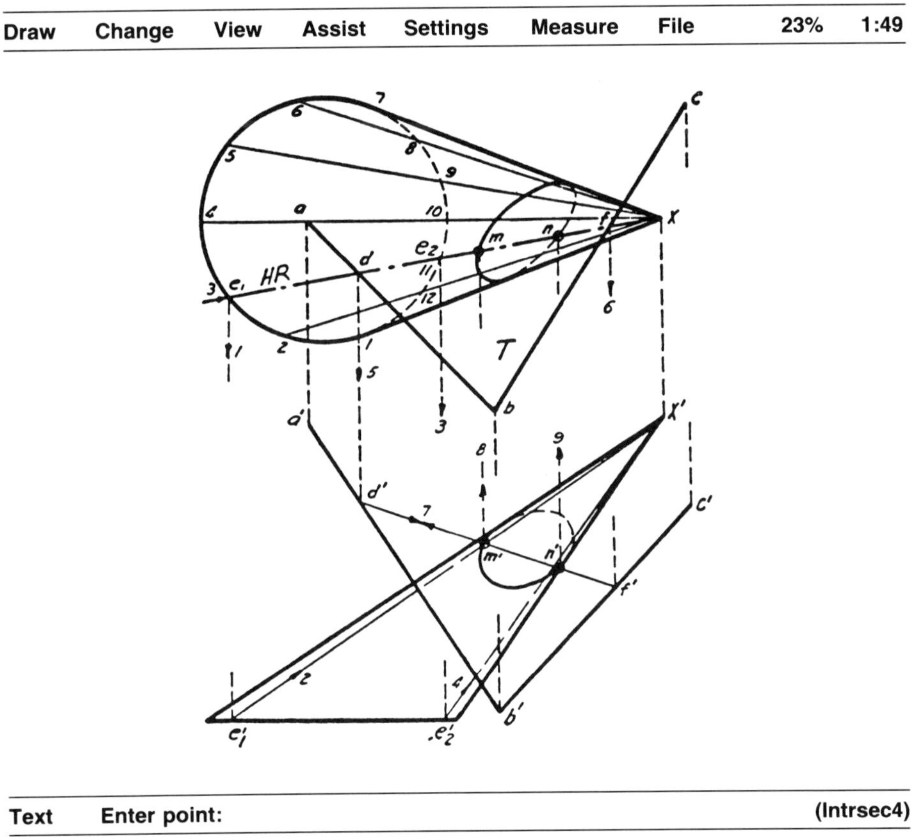

Figure 10.5 Plane of intersection: cone.

Points are displayed at the desired intervals to display the required line of intersection by passing a series of plane surfaces such that they intersect the plane and cone in straight lines. The intersections of each set of lines will be points of the intersection.

10.5 INTERSECTION: CYLINDER

The plane of intersection is displayed in Figure 10.6 by a series of points common to the plane (*T*) and the elements of the cylinder. Points *MN* are displayed as the intersections of lines, which in turn display planes.

Points are displayed at the desired intervals to display the required line of intersection by passing a series of plane surfaces such that they intersect the plane and cylinder in straight lines. The intersections of each set of lines will be points of the intersection.

10.6 INTERSECTION ANGLE: CONE

Case 1 of Figure 10.7 is the analysis, and case 2 is the construction and solution of this common CAE situation. Plane *S* may intersect the cone at any angle through

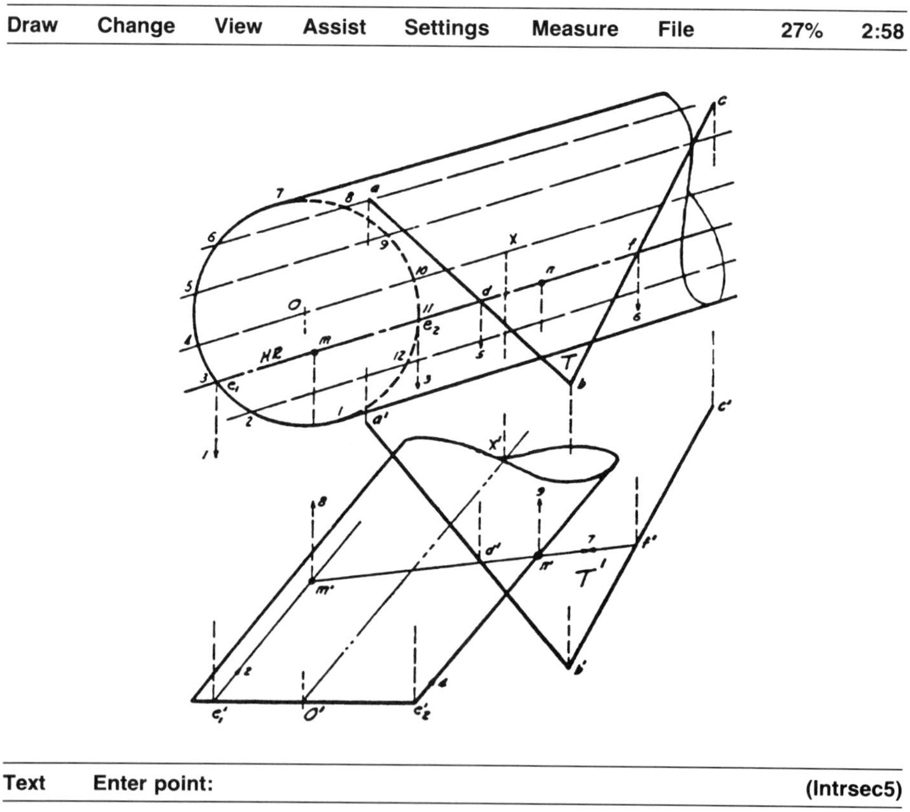

Figure 10.6 Plane of intersection: cylinder.

point A. This angle determines the location of line $H(1)$, $H(2)$ along the base of the cone. If the cutting plane (S) is 90 degrees to plane H in case 1, then the cone is divided into equal halves. As the angle θ changes, less equal parts result as shown in Figure 10.7 (case 2).

10.7 INTERSECTION ANGLE: SPHERE

Figure 10.8, case 1, is the analysis, and case 2 is the construction and solution of this CAE situation. Planes may intersect the sphere as parallel cutting planes (W and X) or at an angle (T). This angle determines the location of line HG along the ellipse of the sphere. If the cutting plane (T) is 90 degrees to axis as shown in case 2, then the sphere is divided into equal halves. As the angle θ changes, elliptical parts result.

10.8 INTERSECTION: CYLINDER AND CONE

Figure 10.9 is the analysis, and Figure 10.10 is the construction and solution of this CAE situation. Cones and cylinders intersect in the three cases shown in Figure 10.9. These are (1) two separate curves, (2) two curves with one or two common points, and (3) single curve. Figure 10.10 is the construction of a type 3, single-curve intersection.

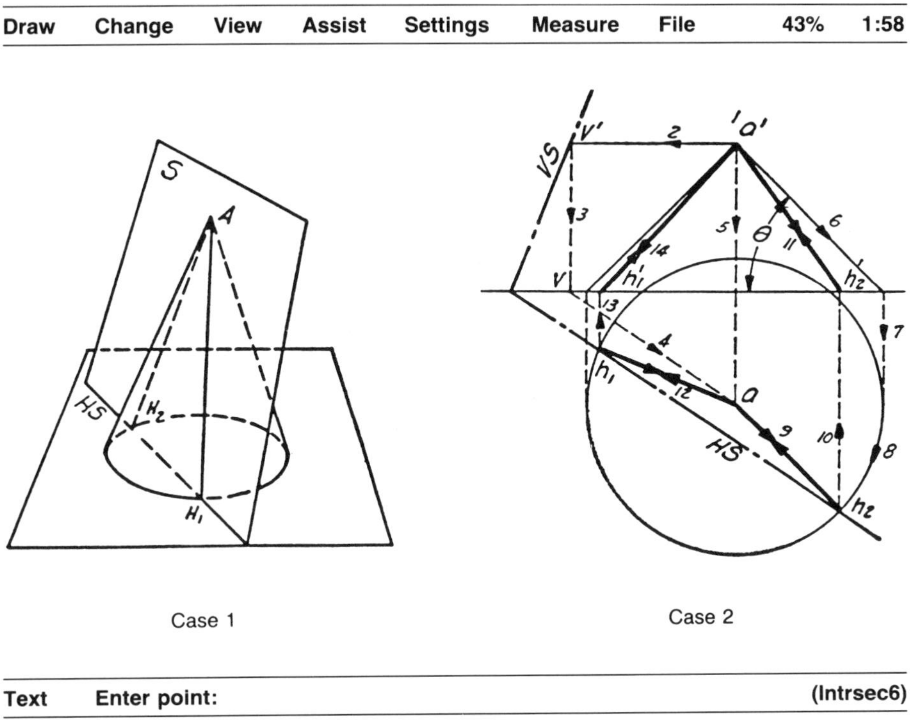

Figure 10.7 Angle of intersection: cone.

Draw **Change** **View** **Assist** **Settings** **Measure** **File** **56%** **2:27**

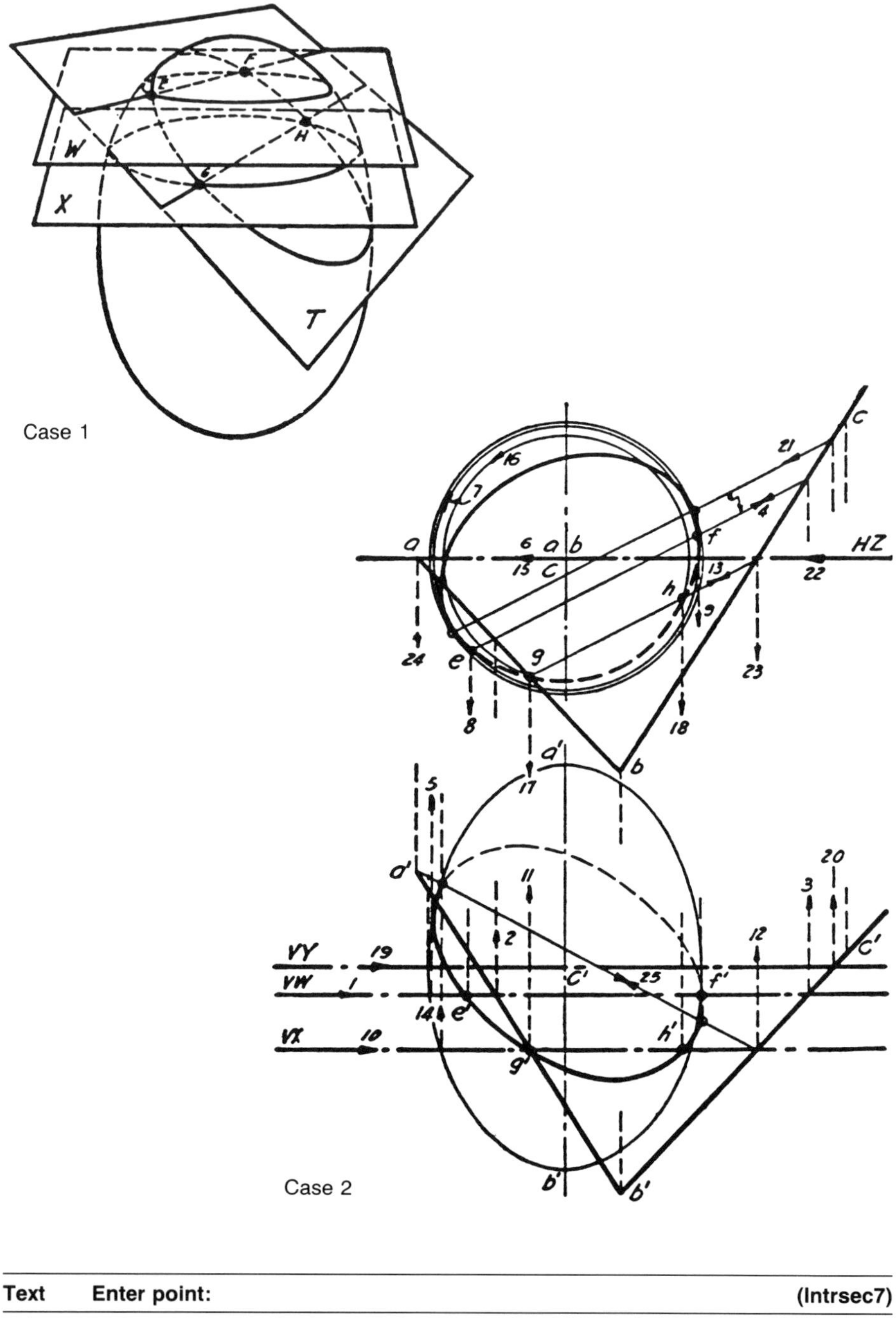

Text **Enter point:** **(Intrsec7)**

Figure 10.8 Angle of intersection: sphere.

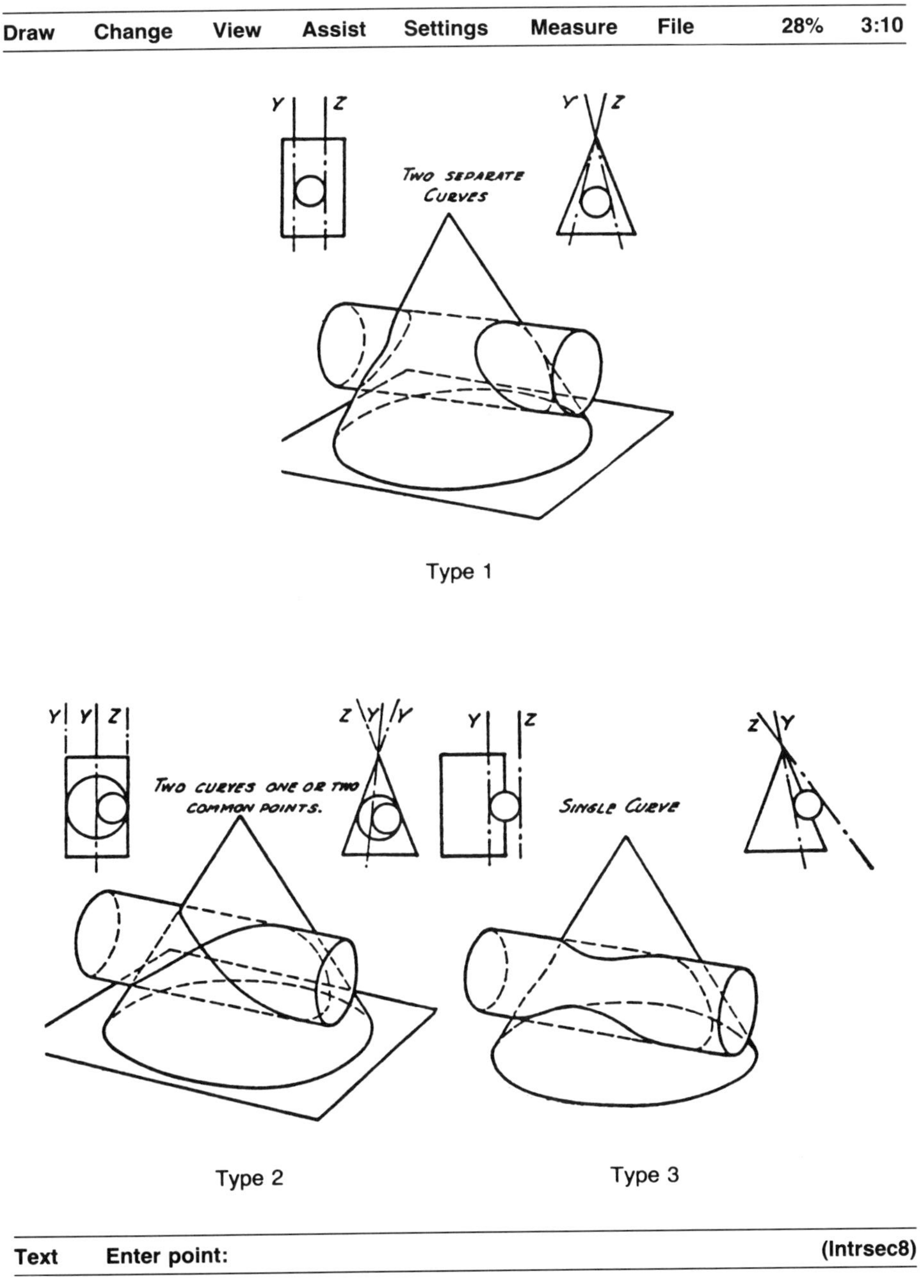

Figure 10.9 Intersection: cone and cylinder.

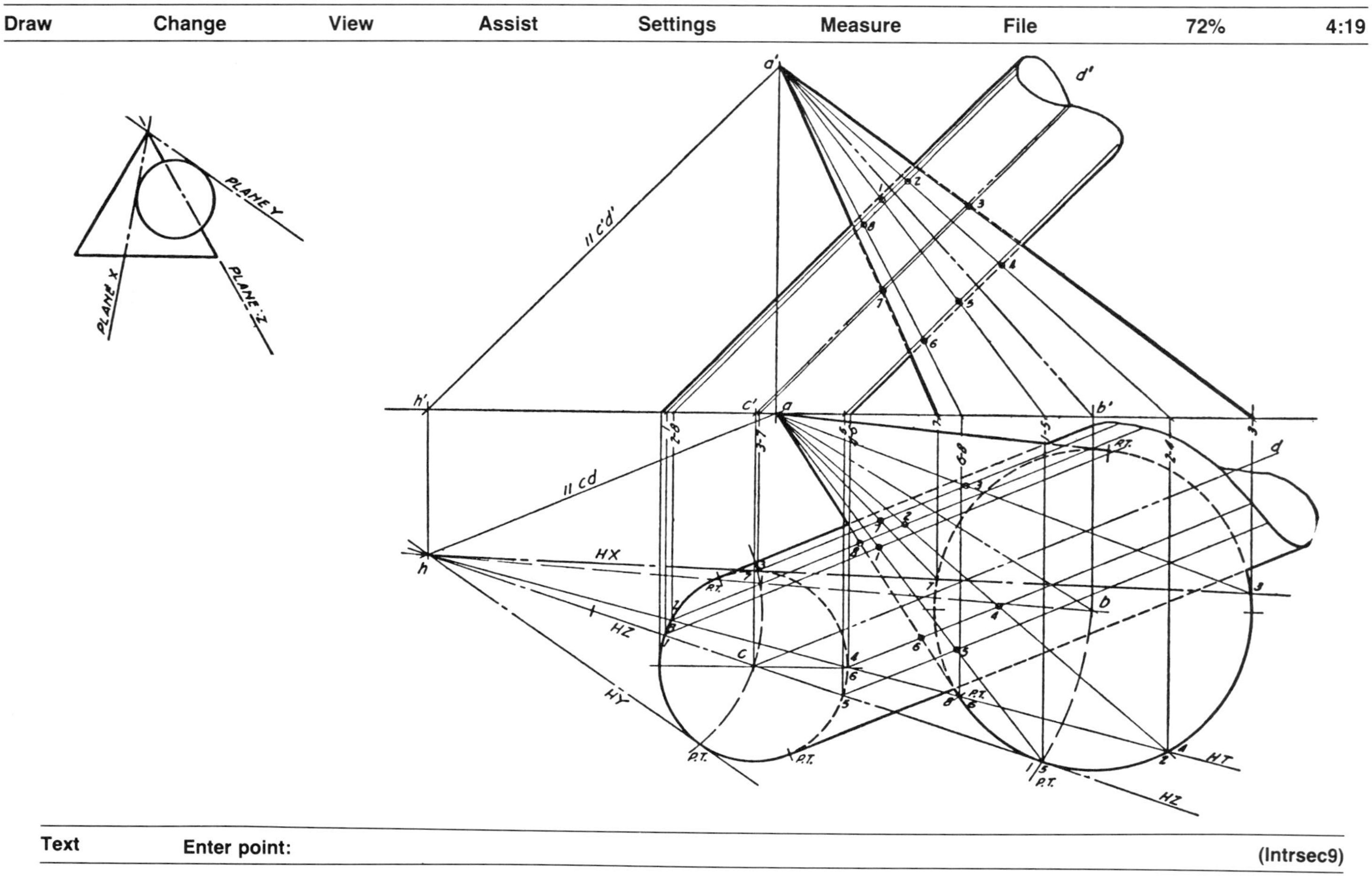

Figure 10.10 Intersection: cone and cylinder, type 3.

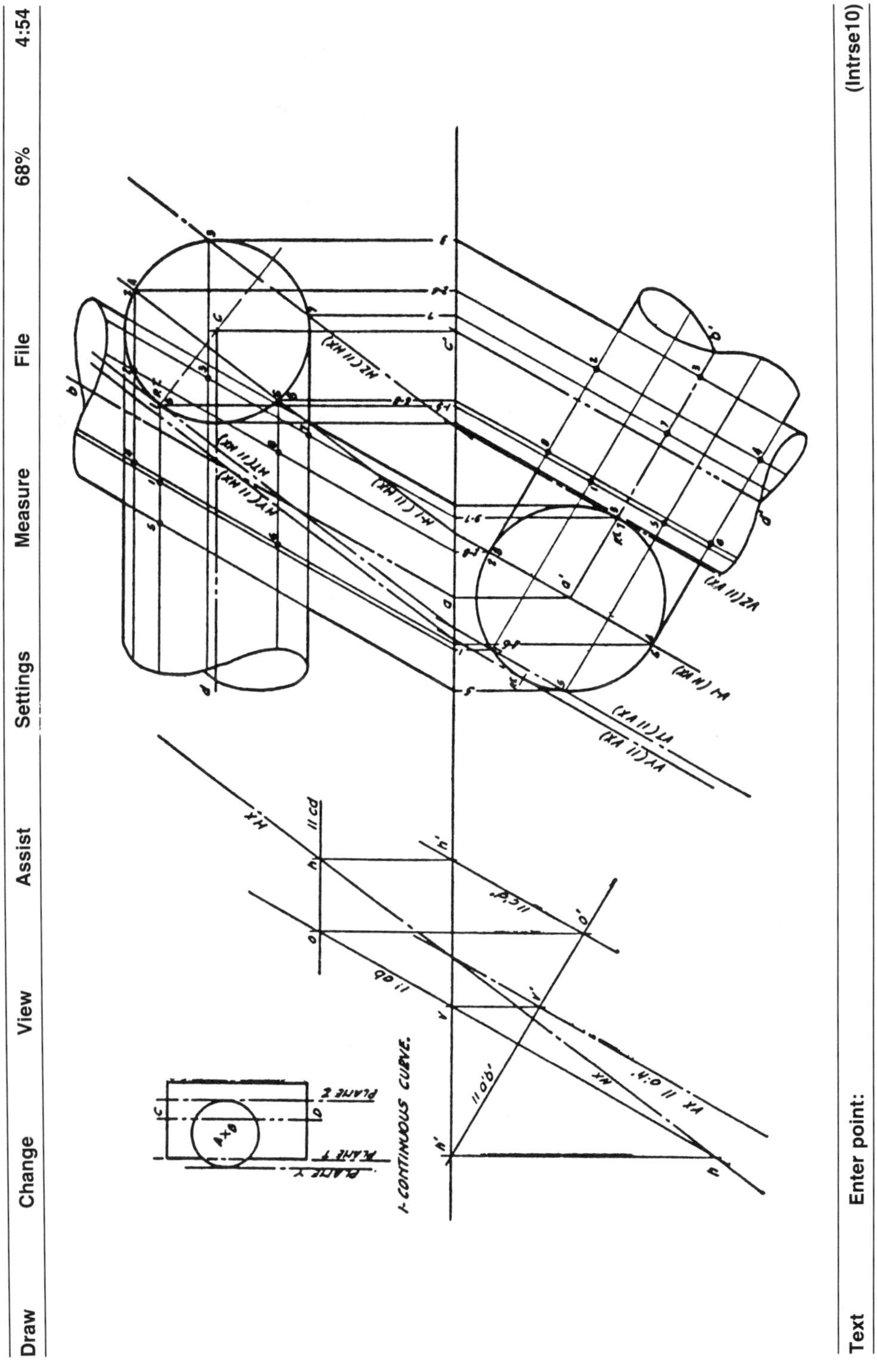

Figure 10.11 Intersection of cylinders, type 3.

10.9 INTERSECTION OF CYLINDERS

Figure 10.9 is the analysis, and Figure 10.11 is the construction and solution of this CAE situation. Cylinders and cylinders intersect in the three cases shown in Figure 10.9. These are (1) two separate curves, (2) two curves with one or two common points, and (3) single curve. Figure 10.11 is the construction of a type 3, single-curve intersection.

10.10 INTERSECTION: EQUAL DIAMETER CYLINDERS

Figure 10.9 is the analysis, and Figure 10.12 is the construction and solution of this CAE situation. Cylinders of equal diameters also intersect in the three cases shown in Figure 10.9. These are (1) two separate curves, (2) two curves with one

Draw Change View Assist Settings Measure File 73% 5:25

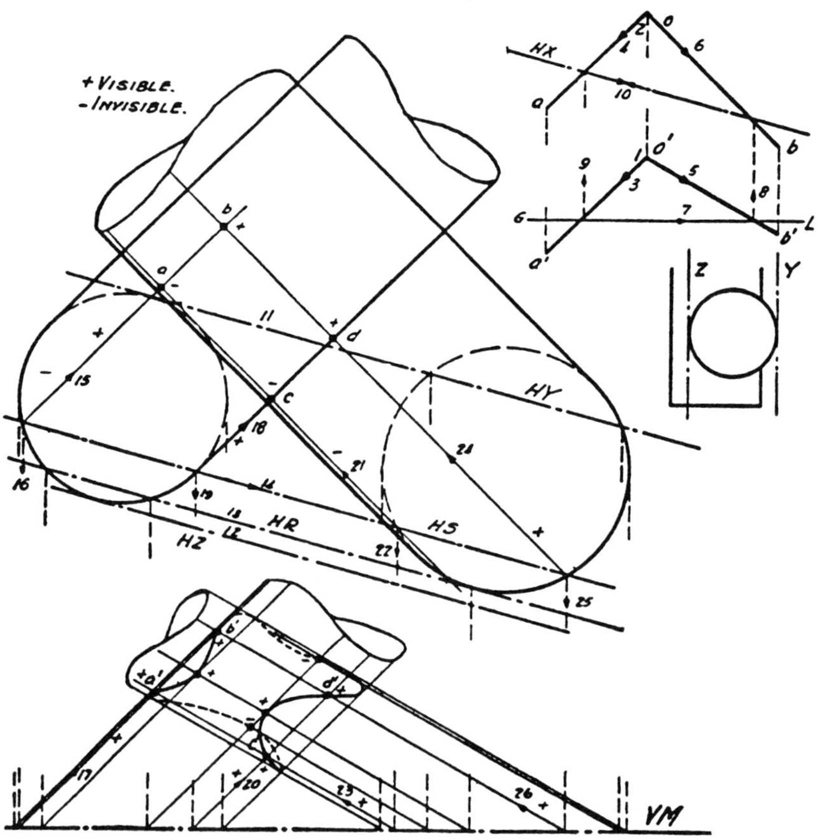

Text Enter point: (Intrse11)

Figure 10.12 Intersection of cylinders: equal diameter.

or two common points, and (3) single curve. Figure 10.12 is the construction of a type 3, single-curve intersection.

10.11 INTERSECTION: UNEQUAL DIAMETER CYLINDERS

Figure 10.9 is the analysis, and Figure 10.13 is the construction and solution of this CAE situation. Cylinders of unequal diameters also intersect in the three cases

Draw Change View Assist Settings Measure File 73% 5:25

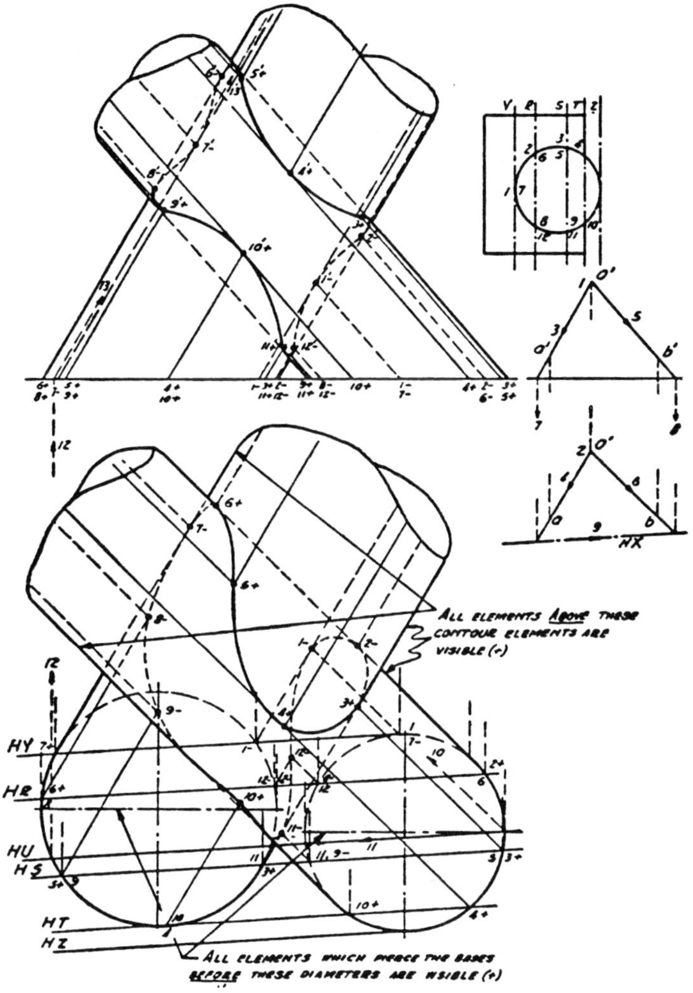

Text Enter point: (Intrse12)

Figure 10.13 Intersection of cylinders: unequal diameter.

shown in Figure 10.9. These are (1) two separate curves, (2) two curves with one or two common points, and (3) single curve. Figure 10.13 is the construction of a type 3, single-curve intersection.

10.12 ANGULAR INTERSECTIONS

Figure 10.9 is the analysis, and Figure 10.14 is the construction and solution of this CAE situation. Cylinders of unequal diameters may intersect at any angle in the three cases shown in Figure 10.9. These are (1) two separate curves, (2) two curves with one or two common points, and (3) single curve. Figure 10.14 is the construction of a type 1, two separate curves intersection.

10.13 INTERSECTION: ANGULAR CONES AND CYLINDERS

Figure 10.9 is the analysis, and Figure 10.15 is the construction and solution of this CAE situation. Cylinders and cones of unequal diameters may intersect at any angle in the three cases shown in Figure 10.9. These are (1) two separate curves, (2) two curves with one or two common points, and (3) single curve. Figure 10.15 is the construction of a type 2, two curves, two common points intersection.

Draw Change View Assist Settings Measure File 59% 6:32

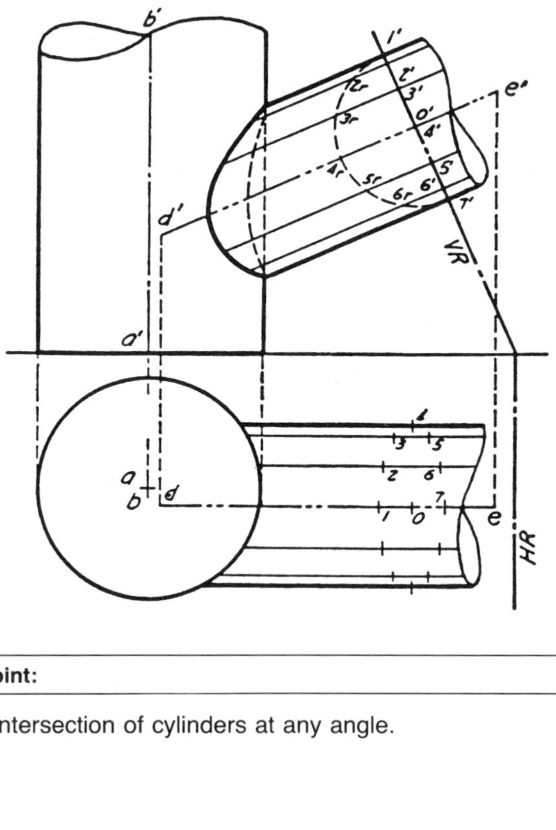

Text Enter point: (Intrse13)

Figure 10.14 Intersection of cylinders at any angle.

10.14 INTERSECTION: ANGULAR CONES AND CONES

Figure 10.9 is the analysis, and Figure 10.16 is the construction and solution of this CAE situation. Cones and cones of unequal diameters may intersect at any angle in the three cases shown in Figure 10.9. These are (1) two separate curves, (2) two curves with one or two common points, and (3) single curve. Figure 10.16 is the construction of a type 2, two curves, two common points intersection.

Draw	Change	View	Assist	Settings	Measure	File	82%	7:39

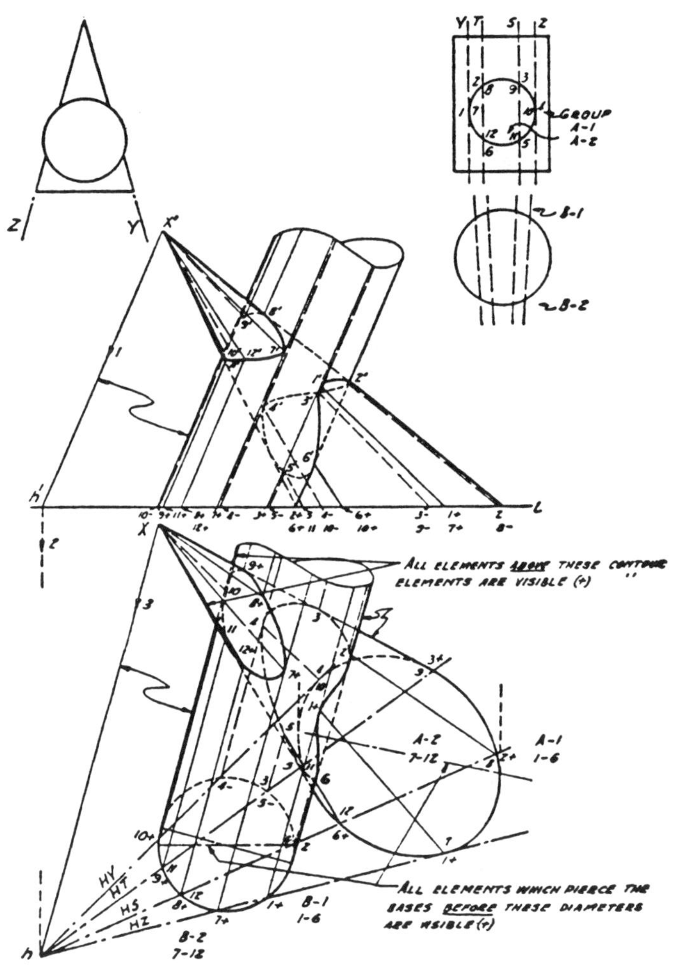

Text Enter point: (Intrse14)

Figure 10.15 Intersection of cylinders and cones at any angle.

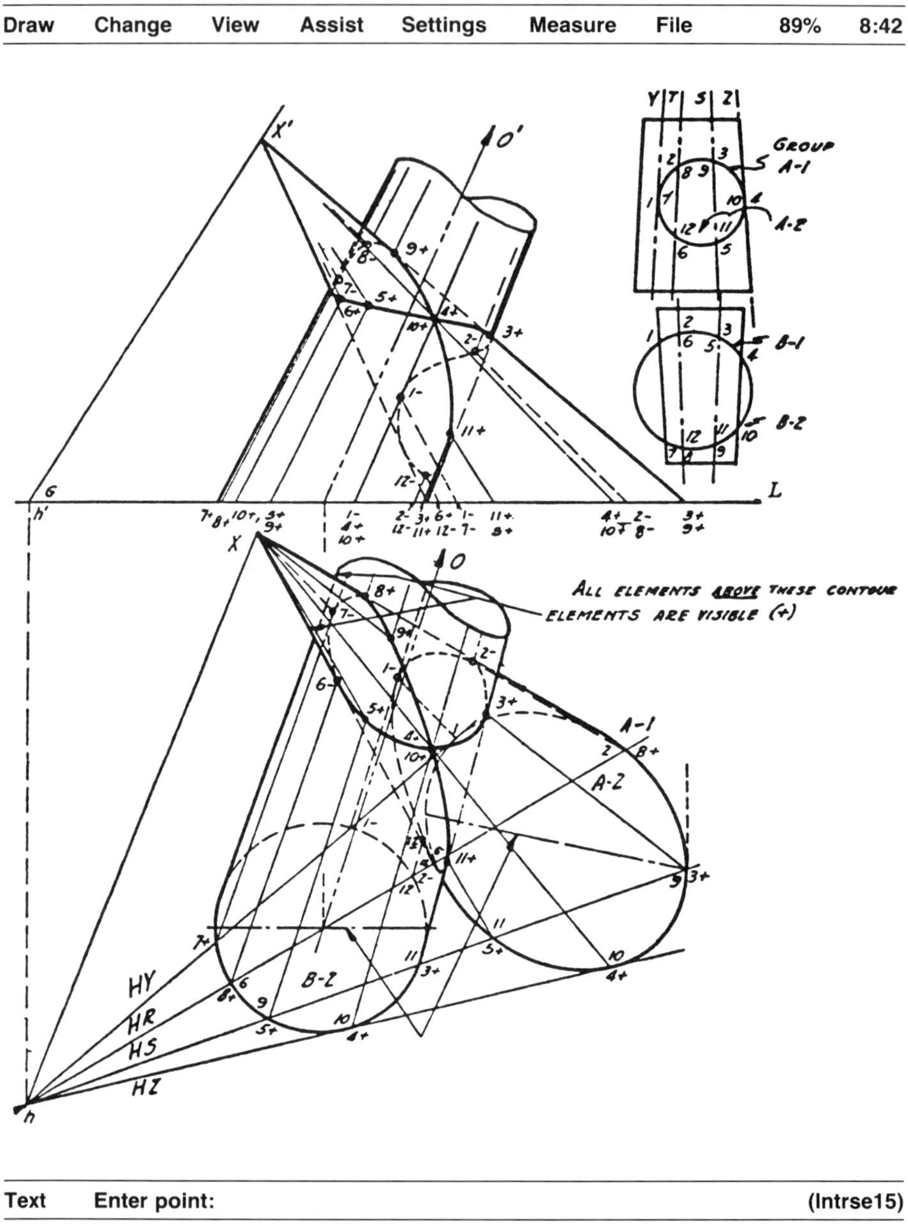

Figure 10.16 Intersection of cones and cones at any angle.

10.15 CHAPTER PROBLEMS

The problems appearing at the end of this chapter have been selected to introduce the basic types of broad, general problems encountered by an engineer or engineering technician using a CADD display system. These problems emphasize the interdisciplinary aspects of constructive geometry, which require a number of related pieces of CADD software. The software used for the production of problems in this chapter can be ordered from Autodesk, Inc.

PROBLEM SET 1 *Cutting plane surfaces*

In each problem display the plane defined.

1.1 Plane *A* (Fig. 10.1 [1])
1.2 Plane *B* (Fig. 10.1 [2])
1.3 Plane *T* (Fig 10.2 [1])
1.4 Plane *T* (Fig. 10.2 [2])
1.5 Plane *T* (Fig. 10.2 [3])
1.6 Plane *T* (Fig. 10.3)
1.7 Plane *T* (Fig. 10.4)
1.8 Plane *T* (Fig. 10.5)
1.9 Plane *T* (Fig. 10.6)
1.10 Plane *S* (Fig. 10.7)

PROBLEM SET 2 *Cylinder displays*

In each problem display the cylinder defined.

2.1 *R* (Fig. 10.6)
2.2 *Y* (Fig. 10.9)
2.3 *C* (Fig. 10.10)
2.4 *B* (Fig. 10.11)
2.5 *P* (Fig. 10.12)
2.6 *Y* (Fig. 10.12)
2.7 *X* (Fig. 10.13)
2.8 *R* (Fig. 10.14)
2.9 *H* (Fig. 10.14)
2.10 *X* (Fig. 10.15)

PROBLEM SET 3 *Cone displays*

In each problem display the cone defined.

3.1 *X* (Fig. 10.3)
3.2 *X* (Fig. 10.4)
3.3 *A* (Fig. 10.5)
3.5 *A* (Fig. 10.7 [1])
3.6 *Z* (Fig. 10.7 [2])
3.7 *A* (Fig. 10.9)
3.8 *O* (Fig. 10.10)
3.9 *C* (Fig. 10.15)
3.10 *A* (Fig. 10.16)

11 Development of Solid Surfaces

This chapter introduces the CADD development of a plane surface into a solid. Prior to this chapter the relationship has been the intersection of any plane with a solid or solid intersections. The series of display solids shown in this chapter consists of the linking of planes to form solids. Each of the display situations is a section heading; under each heading is the CADD construction required for the display.

The development of a simple solid (cylinder or cone) is the surface of that solid rolled out on a plane. The plane is formed by passing tangent to the solid through one of its elements, and the surface is then split along any element and rolled out on this tangent plane (see Chapter 9).

11.1 LAYOUT METHOD

The layout of the conduit shown in Figure 11.1 is made up of at least 22 separate solids comprising sections of cylinders and cones. Each was displayed from a set of tangent planes as described in Chapter 9 of this text. The necessary provision for fitting one solid to another, end to end or side to side, was completed to form Figure 11.1.

It is the purpose of this chapter to cover the details of solid layout. The process of producing tangent planes equivalent to the total surface of a solid is called the *layout method*.

11.2 APPROXIMATION TECHNIQUES

The process of producing a layout equivalent to the surface of a cylinder or a cone and any lines on those surfaces may be displayed simpler if the cone is regarded as satisfactorily approximated for engineering purposes by an inscribed pyramid (large number of faces), and the cylinder approximated by an inscribed prism with a large number of faces.

The number of faces in each case needs to be large in order that a layout may satisfactorily approximate the desired curved surface of the solid. Developments are frequently worked out in two or more parts because of the sizes of the tangent planes required. It is better to regard the solid in terms of an inscribed prism or pyramid with a number of sides divisible by two, three, or four.

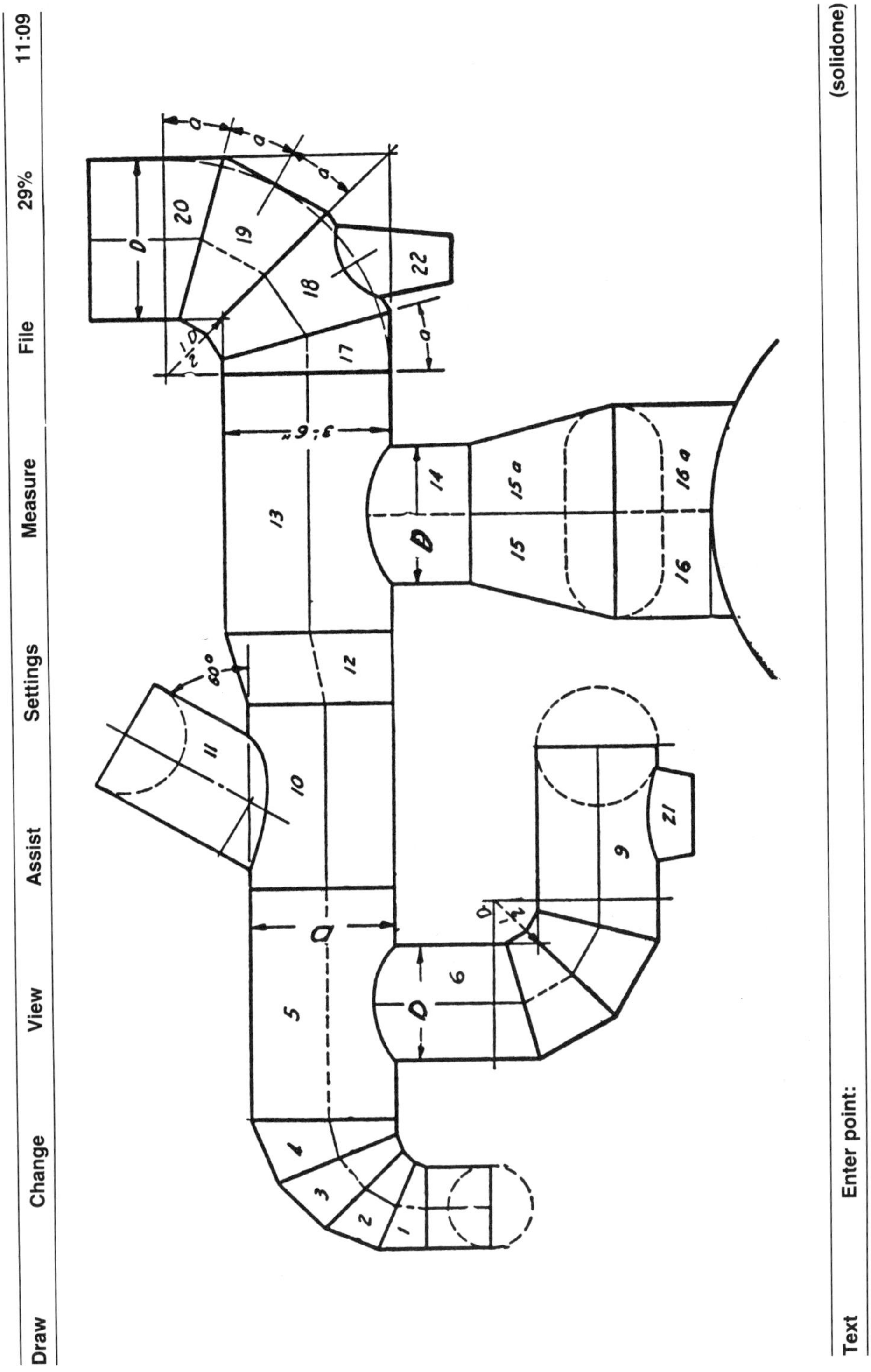

Figure 11.1 Compound solid display.

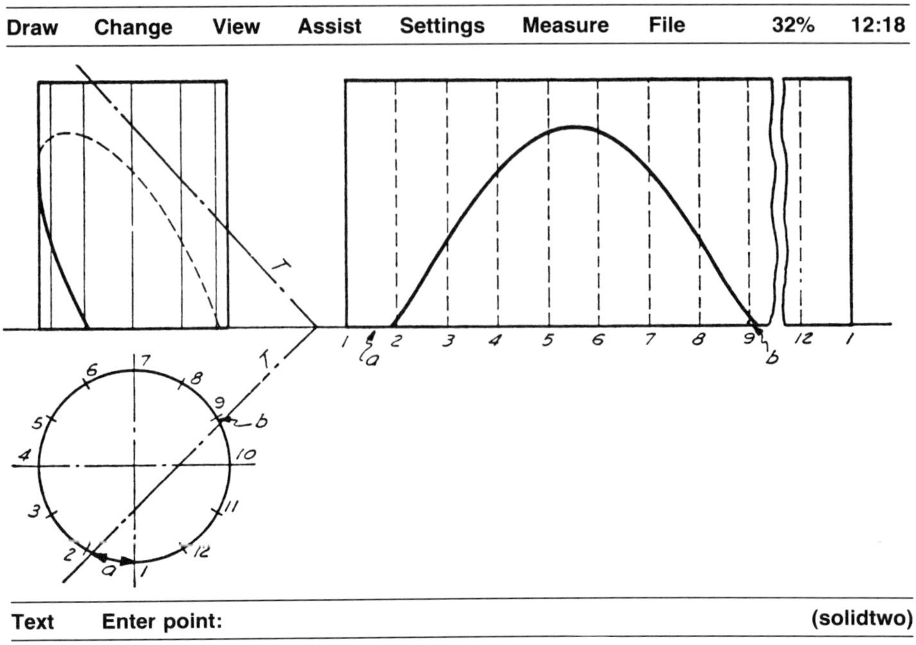

Figure 11.2 Cylinder development.

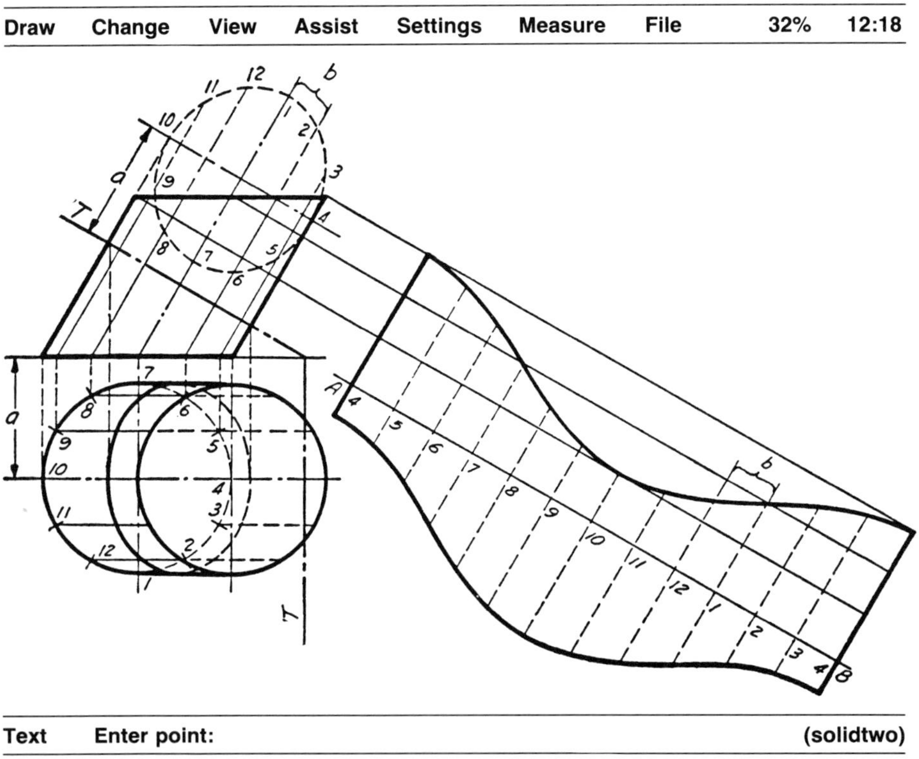

Figure 11.3 Oblique cylinder development.

11.3 DEVELOPMENT OF A CYLINDER

The development of the cylinder begins with an end view divided into 12 faces as shown in Figure 11.2. These divisions are then projected into the adjacent view, and a development is displayed to the right of this view. The chord of one of the 12 arcs of the end view is laid off 12 times on a horizontal line 1-12-1. If the cylinder is to cut, e.g., cutting plane line *T* in Figure 11.2, this may also be indicated as shown in Chapter 10 of this text. The surface of intersection is shown in both orthographic views and in the development.

11.4 OBLIQUE CYLINDER DEVELOPMENT

The development of the oblique cylinder begins at the base as it appears as an ellipse. These 12 divisions are then projected into the adjacent view, and a development is displayed parallel to the axis. The chord (*b*) of one of the 12 arcs of the base is laid off 12 times on a perpendicular line 4-12-4. If the cylinder is to be cut, e.g., cutting plane line *T* in Figure 11.3, this may also be indicated as shown in

Draw	Change	View	Assist	Settings	Measure	File	43%	1:19

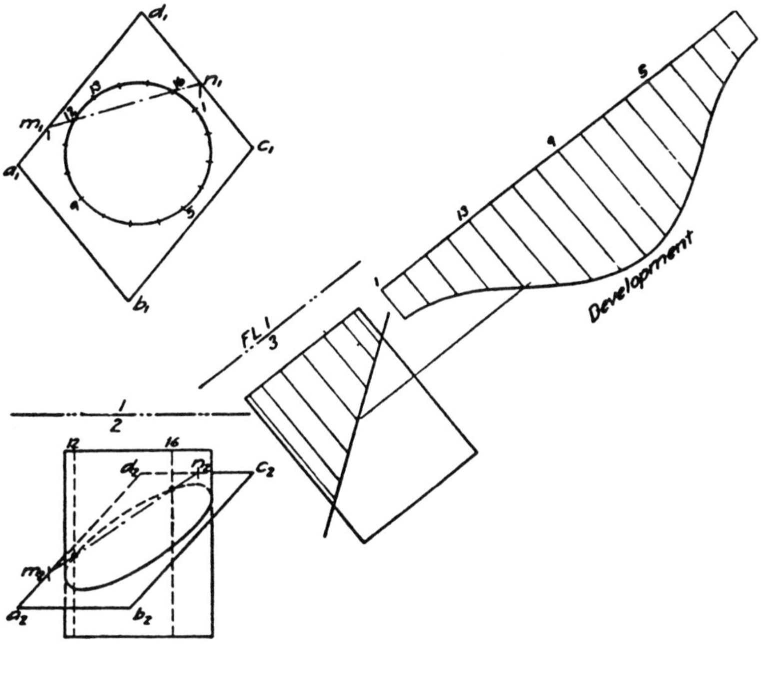

Text	Enter point:	(solidfor)

Figure 11.4 Right cylinder development.

Chapter 10 of this text. The surface of intersection is shown in both orthographic views and in the development.

11.5 RIGHT CYLINDER DEVELOPMENT

The development of the right cylinder begins at the base as it appears as a circle. These 16 divisions are then projected into the adjacent view, and a development is displayed parallel to the axis. The chord of one of the 16 arcs of the base is laid off 16 times on a perpendicular line 1-16-1. If the cylinder is to be cut, e.g., cutting plane line *MN* in Figure 11.4, this may also be indicated as shown in Chapter 10 of this text. The surface of intersection is shown in both orthographic views and in the development.

11.6 CONE DEVELOPMENT

The development of the right cone begins at the base as it appears as a circle. These 16 divisions are then projected into the adjacent view, and a development is displayed parallel to an element. The chord of one of the 16 arcs of the base is laid

Draw **Change** **View** **Assist** **Settings** **Measure** **File** **38%** **2:35**

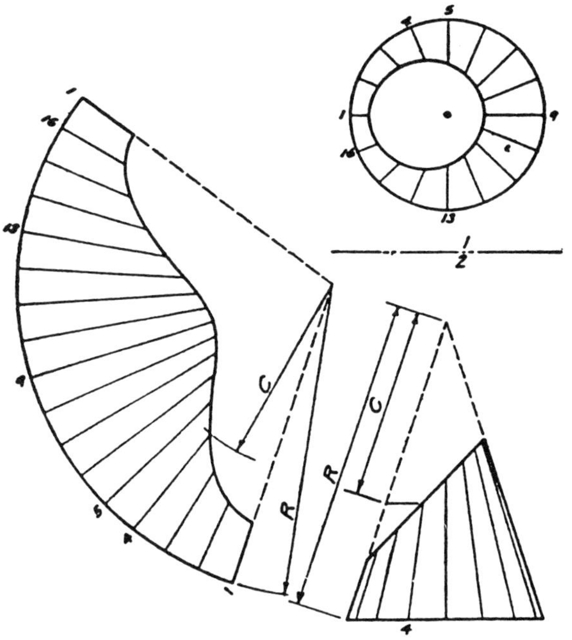

Text **Enter point:** **(solidfiv)**

Figure 11.5 Cone development.

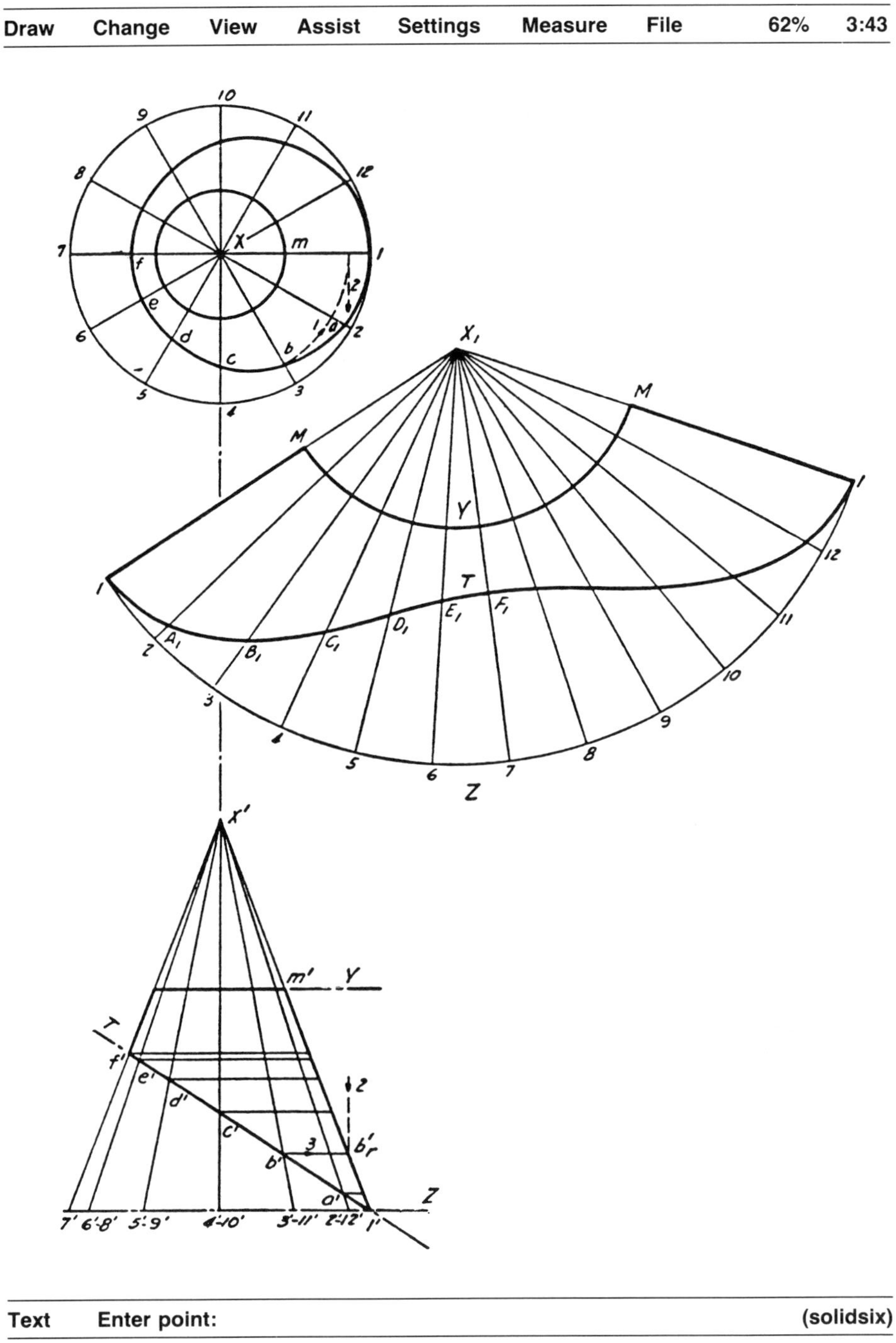

Figure 11.6 Truncated cone development.

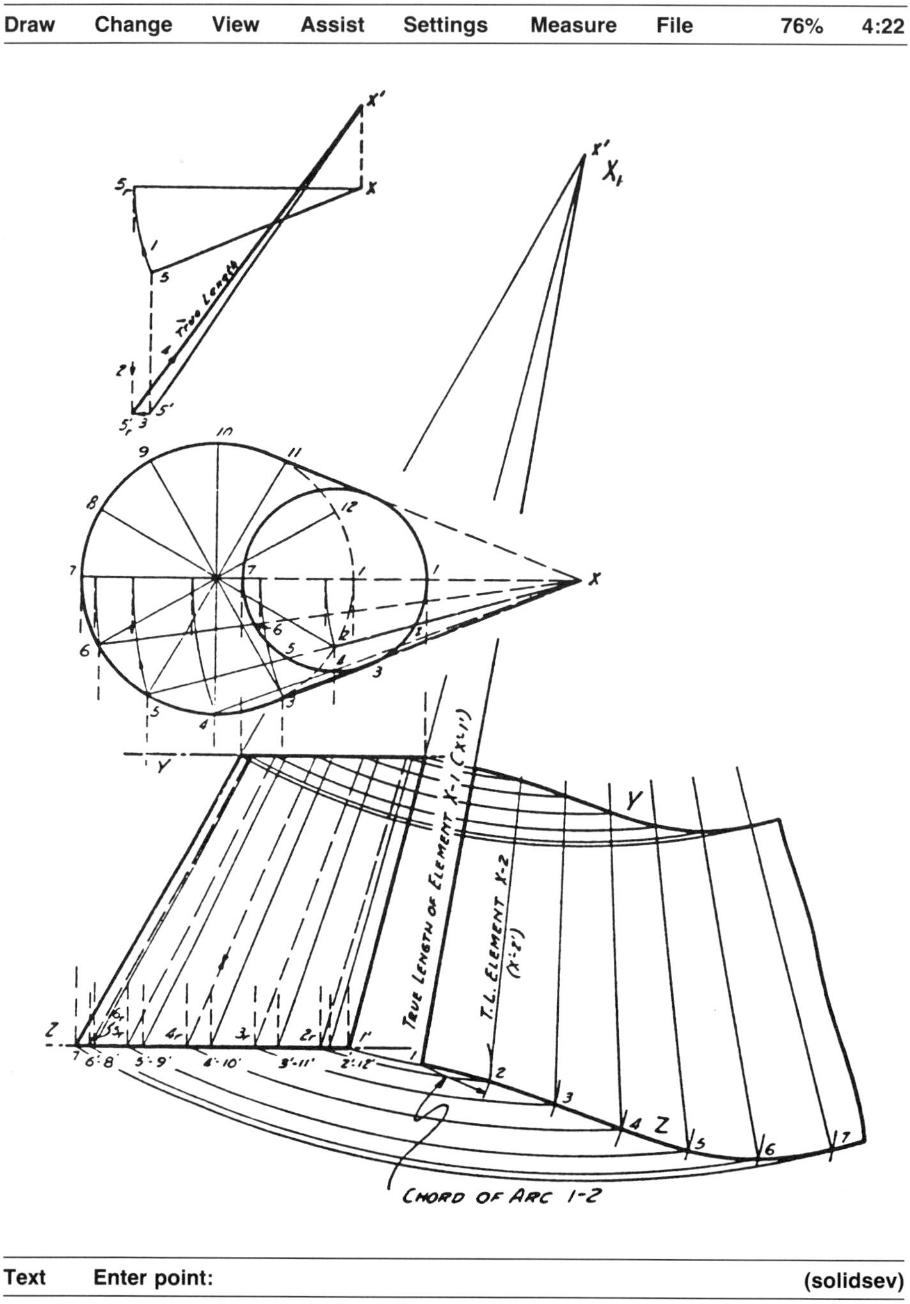

Figure 11.7 Oblique cone development.

off 16 times on a circular arc 1-16-1. If the cone is to be cut, e.g., cutting plane line *C* in Figure 11.5, this may also be indicated as shown in Chapter 10 of this text. The surface of intersection is shown in both orthographic views and in the development.

11.7 TRUNCATED CONE DEVELOPMENT

The development of the truncated cone begins at the base and along the line of truncation (cutting plane *T*). These 12 divisions are then projected into the adjacent view and placed upon the base circle. The line of truncation is divided into 12 parts also (a through f and a' through f'). These divisions are then projected into the adjacent view. A development is displayed at any convenient screen location. The chord of one of the 12 arcs of the base is laid off 12 times on a circular arc 1-12-1. Now the cutting data (a through f and a' through f') is displayed relative to circular arc *Z*. If this cone is to be cut, e.g., cutting plane line *M* in Figure 11.6, this may also be indicated as shown in Chapter 10 of this text. The surface of intersection is shown in both orthographic views and in the development.

11.8 OBLIQUE CONE DEVELOPMENT

The development of the oblique cone begins at the base as it appears as an ellipse. These 12 divisions are then projected into the adjacent view and revolved for true length. A development is displayed at any convenient screen location. The chord of one of the 12 arcs of the base is laid off 12 times on a horizontal line *Z* in Figure 11.7. If the cone is to be cut, e.g., cutting plane line *Y* in Figure 11.7, this may also be indicated as shown in Chapter 10 of this text. The surface of intersection is shown in both orthographic views and in the development.

11.9 CHAPTER PROBLEMS

The problems appearing at the end of this chapter have been selected to introduce the basic types of broad, general problems encountered by an engineer or engineering technician using a CADD display system. These problems emphasize the interdisciplinary aspects of constructive geometry, which require a number of related pieces of CADD software. The software used for the production of problems in this chapter can be ordered from Autodesk, Inc.

PROBLEM SET 1 *Development of surfaces*

In each problem display the surface defined.

1.1 Plane 1 (Fig. 11.1)
1.2 Plane 2 (Fig. 11.1)
1.3 Plane 3 (Fig. 11.1)
1.4 Plane 4 (Fig. 11.1)
1.5 Plane 5 (Fig. 11.1)
1.6 Plane 6 (Fig. 11.1)
1.7 Plane 7 (Fig. 11.1)

1.8 Plane 8 (Fig. 11.1)
1.9 Plane 9 (Fig. 11.1)
1.10 Plane 10 (Fig. 11.1)

PROBLEM SET 2 *Cylinder displays*

In each problem display the cylinder defined.

2.1 5 (Fig. 11.1)
2.2 10 (Fig. 11.1)
2.3 *T* (Fig. 11.2)
2.4 *T* (Fig. 11.3)
2.5 *C* (Fig. 11.4)
2.6 11 (Fig. 11.1)
2.7 13 (Fig. 11.1)
2.8 14 (Fig. 11.1)
2.9 9 (Fig. 11.1)
2.10 *D* (Fig. 11.1)

PROBLEM SET 3 *Development displays*

In each problem display the cone defined.

3.1 *T* (Fig. 11.2)
3.2 *H* (Fig. 11.2)
3.3 *A* (Fig. 11.3)
3.4 *T* (Fig. 11.3)
3.5 *A* (Fig. 11.4)
3.6 *Z* (Fig. 11.4)
3.7 *C* (Fig. 11.5)
3.8 *O* (Fig. 11.6)
3.9 *C* (Fig. 11.6)
3.10 *A* (Fig. 11.7)

12 Descriptive Geometry Applications

This chapter is the result of a search of the *Compendex Plus* database using the DIALOG Information Retrieval Service (DIRS). This database provides coverage of the world's significant engineering and technology literature. *Compendex Plus* database is produced by Information Services, Inc., located at 3460 Hillview Avenue in Palo Alto, CA 94304. This company also produces *Engineering Index*. Each record in the database is a reference to a journal article, technical report, engineering society publication, textbook, or conference proceedings and includes a concise abstract describing the descriptive geometry application. Prior to January 1988, this database existed as two databases: *Compendex* (file 8) and *Ei Engineering Meetings* (file 165). These two files were merged to facilitate one-step searching.

12.1 SAMPLE APPLICATIONS

The display screen shown in Figure 12.1 represents the first reference stored in the database. The line across the top of the screen indicates that you have accessed the DIALOG file 8, which is the Compendex Plus database for the years 1977 through September 1989. The second line is a database code number, while the remainder of the lines are labeled: title, author, affiliations, sponsors, sources, dates, languages, document types, and abstract. Each of remaining sections of this chapter are selected applications for descriptive or constructive geometry.

12.2

Towards generation theory of freshman graphic reading error. I. Basic knowledge and simple learning model for graphic reading
Kajiyama, K.; Ezaki, T.; Irie, K.; Nagatomo, K.
Author Affil: Fukuoka Univ., Japan
Source: Fukuoka Univ. Rev. Technol. Sci. (Japan) no. 41, pp.: 161-70
Publication Year: Sept. 1988
CODEN: FDKSDF ISSN: 0285-2799
Treatment: THEORETICAL;
Document Type: JOURNAL PAPER
Languages: Japanese
(10 Refs)

Abstract: The authors' CAI system for visual enhancement in descriptive geometry has only a basic function of solution error checking. This paper describes an error graphic model for expanding to an intelligent CAI system. Basic knowledge of

DIALOG File 8: COMPENDEX PLUS—77-89 (SEP Copr. Engineering Info Inc. 1989)

02791390 E.I. Monthly No: EI8909090511

Title: Use of descriptive geometry to determine the position and orientation of a revolute joint for spatial motion of two points in a rigid body.

Author: Beard, J. E.

Corporate Source: Louisiana State Univ, Baton Rouge, LA, USA

Conference Title: Trends and Developments in Mechanisms, Machines, and Robotics—1988 Presented at the 1988 ASME Design Technology Conferences 20th Biennial Mechanisms Conference

Conference Location: Kissimmee, FL, USA Conference Date: 1988 Sep 25-28

Sponsor: ASME, Design Engineering Div, New York, NY, USA

E.I. Conference No.: 12265

Source: American Society of Mechanical Engineers, Design Engineering Division (Publication) DE v 15-1 PTI. Publ by ASME, New York, NY, USA. p 441-444

Publication Year: 1988

CODEN: AMEDEH

Language: English

Document Type: PA; (Conference Paper) Treatment: T; (Theoretical)

Journal Announcement: 8909

Abstract: In order to design a mechanism with a single revolute joint, where the desired positions of two separate points, in a rigid body, are specified in the initial and final positions in space, the revolute matrix, left bracket R right bracket, a special case of the 4 multiplied by 4 screw matrix left bracket S right bracket, can be used to determine the orientation of the axis and a point through which this axis passes. This paper presents a method of using descriptive geometry to find the revolute joint in space where the method of solution may help gain insight into kinematic synthesis. (Edited author abstract) 5 Refs.

Figure 12.1 Sample record.

graph reading is investigated from many textbooks and the analysis of reading experiments using computers. From these investigations the possibility of systematic explanation for beginner's error is found.

12.3

Application of multidimensional descriptive geometry to phase diagram representation of multicomponent metal system

Zhou Jiyi; Xie Shenjian
Author Affil: Dept. of Precision Instrum. & Mechanology, Tsinghua Univ., China

Source: J. Tsinghua Univ. (China) vol.28, no.5, pp.: 62-71
Publication Year: 1988
CODEN: CHHPAH ISSN: 0577-9189
Treatment: THEORETICAL;
Document Type: JOURNAL PAPER
Languages: Chinese
(8 Refs)

Abstract: The authors present a new method to represent the components and constitution of four and five component metal systems using multiview projection drawing on the basis of multidimensional descriptive geometry. Based on such drawing, it is possible to predict fusion temperatures and crystallization lines of any component metal in a multicomponent metal system, and on the basis of experimentation, to map out the probable phase diagram of multicomponent system by means of known phase diagrams of lower component systems. The experimental and research work in making such phase diagrams will thus be simplified.

12.4

Proceedings of the Third International Conference on Engineering Graphics and Descriptive Geometry
Slaby, S.M.; Stachel, H. (Editors)
Sponsor: ASEE; Univ. Graz; Univ. Innsbruck; Univ. Leoben; Univ. Vienna
Publication Year: 1988
Conference Information:
Proceedings of the Third International Conference on Engineering Graphics and Descriptive Geometry Date: 11-16 July 1988 Location: Vienna, Austria
Publ: Tech. Univ. Vienna, Vienna, Austria. 2 vol. (ix+335+vii+349) pp.
Treatment: PRACTICAL;
Document Type: CONFERENCE PROCEEDINGS
Languages: English

A descriptive geometric approach to constructions

Abstract: A method to find the radius and centre of a circle in a plane is developed. It discriminates unambiguously among up to eight possible solutions. Any necessary and sufficient combination of constraining radius, centre line(s), tangent line(s) and/or tangent circle(s) may be used to specify the construction. Solutions proceed as the unique intersection among three planes, if there are no tangent circles. Otherwise there are two; the piercing points of the dihedral line of two planes and the surface of a cone. Selection between these can be based on difference in radius or of centre location.

Solving the workspace problem of a robot on the basis of graphical and analytical method

Abstract: The workspace of a given robot is a point set within which every point can be reached by a reference point on the robot hand. One of the most important problems in robot design is to determine the shape of the workspace. The authors present a graphical-analytical method, which can determine quickly and intuitionally the shape of the workspace on an intersection plane. They discuss the three-intersection-link chopstick robot. The graphical-analytical method is more convenient, more visualized and easier to understand than the others.

The Monge method applied to determination of angles in the four-dimensional Euclidean space

Abstract: The paper looks at how to apply the Monge method to the determination of angles between non-intersecting lines and planes, between lines and hyperplanes, between two planes intersecting at one point, between planes and hyperplanes and between two hyperplanes in four-dimensional Euclidean space. It also gives diagrams for determining angles between two planes intersected at one point.

A teaching technique for the teaching of descriptive geometry

Abstract: A teaching technique has been devised for teaching descriptive geometry that in the opinion of the author will: keep the students up-to-date in the course material; improve class attendance; reduce tardiness to almost zero; provide an automatic attendance record; sharpen the student's ability to analyze under pressure; provide the instructor with a continuous feedback as to the student's mastery of the subject matter; distribute the load of testing over the entire semester; enable testing instruments to be evaluated very rapidly; and provide the student with a continuous update of his progress in the course. The heart of this teaching technique is the quiz; there is no need for any drawing on the part of the student. This system is described.

Descriptive geometry–the science of visualization

Abstract: The author states that geometry is the basis of mathematics since points, lines, planes and solids existed in nature many milleniums before symbols (mathematics) were derived, as a means to describe and manipulate the geometry and geometric relationships which exist in the real world. He discusses descriptive geometry.

Geometrical modelling of new tore and torelike forms

Abstract: All kinds of kinematic tore surfaces formed by rotating circles differently disposed relative to the axis are examined. A theorem of possibility to draw two tore surfaces through two congruent circumferences disposed in space is formulated and proved. The author gives a new approach to the problem of structural synthesis of lever gears with rotating couples being modelled by tores. He investigates a geometric structure problem of tore circumferences of general and particular types.

Computer-aided constructive geometry–do it 3-D!

Abstract: 2-D and 3-D computer-aided constructive geometry (CACG) is done in a quite natural way much as 2-D basic euclidean geometry. With a computer and a plotter (and good software) otherwise unobtainably accurate drawings can be produced fast and easily without needing practical drawing skills. Constructions are not confined to a drawing plane. CACG can solve 3-D problems with 3-D operations immediately 3-D. That is the way geometry is applied in industry today and how it should be taught to engineering students in universities.

Mongean, affine and projective models of three- (or four-) dimensional spaces

Abstract: Reviews existing definitions of descriptive geometry and examines the possibility of defining it in accordance with Klein's criteria (1921). The Mongean drawing is treated as a model of three- (or four-) dimensional space with the undefined abstract concepts of point and line conventionally denoted by a pair of dots (or small circles etc.) and a pair of pencil- (or ink-) marks drawn along a straightedge respectively. The transition between the traditional Euclidean and the Mongean models ('Mongean transformation') is regarded as a one-one identity transformation and the commonly used methods of 'auxiliary views,' 'revolution' and 'parallel translation' are shown to be motions ('isometries'). The paper introduces models of affine and projective spaces, describes affine and projective transformations of three-dimensional spaces into themselves and attempts to treat the constructions commonly used in descriptive geometry as their particular cases. Brief comments on some difficulties in solving descriptive geometry problems are also included.

Descriptive solid geometry (DSG). A new approach to solids modeling on computers

Abstract: Computerized analysis of three-view orthographic drawings represents the beginnings of a solid modeling system, descriptive solid geometry (DSG), based on the laws of descriptive geometry. DSG applies conventional concepts of descriptive geometry and graph theory to interpret orthographic drawings directly as a solid model. This technique allows the mechanical designer to utilize the powerful design and modeling tools of orthographic drawings, while retaining the power and utility of conventional solid models.

Solid object input system by conformal projection method

Abstract: Proposes an input system for 3D objects using the projective plane with the conformal projection method, in which one uses only the display and a two-dimensional digitizer or tablet. These are provided with the usual computer graphics system. The authors introduce a new data input system for complex solids using the projective plane and show both wireframe and shaded solid objects obtained from the system.

Tipped auxiliary and direct projection

Abstract: Double auxiliary projection makes direct relations to the start unknown. The author deals with its improvement. It is performed by means of a tipped auxiliary plane of projection. Orthographic projection to such a plane is obtainable with the help of Wunderlich's divider-construction for axonometry (1953). The author arranges this projection directly to the plan by rebating to HP as the drawing plane. Both projections together organize an extended Monge's system, while two adjacent usual projections organize a proper one. If this is tipped as a whole, a tipped Monge's system comes out. Similarly, one is guided further to a coordinated triple set of tipped auxiliary projections. This set reproduce the well-known configuration of Schmid's cutting ray method (1922). This study gives the tipped auxiliary projection and the direct projection some new interpretations of the reflecting line for the ray.

Traditional descriptive geometry on the personal computer with AutoCAD

Abstract: Computer Aided Design and Drafting (CADD) has changed the way engineers, architects, scientists and contractors conduct their business. To be able to use the modern tools it is still necessary to understand/exercise fundamentals of drafting and descriptive geometry, since they are essential in visualizing complex objects presented in a 2-D drawing.

Plane kinematic geometry and computer graphics

Abstract: The author uses formulas of plane kinematics in a program, which enables one to draw the trajectory of a point, its tangent line and osculating circle and the fixed and moving polhode of a given planar motion.

Application of descriptive geometry in the geometry of robot motion

Abstract: Studies geometric properties of the motion of the effector of a robot manipulator. The author develops the basic properties of a p-dimensional Euclidean space motion as a submanifold of the homogeneous space of all positions of the moving space in the fixed one. He defines the robot manipulator as the product of p1-parametric subgroups of the Lie group of all congruences of the Euclidean space. After that he derives the system of differential equations, which characterizes the robot motion as a p-dimensional space motion and shows how some of its invariant geometrical objects can be constructed by means of descriptive geometry.

Applied descriptive geometry problems

Abstract: Presents geometrical problems from the fields of natural and technical sciences. The benefit of descriptive geometry is twofold: on the one hand it aids in analyzing and reasoning, special views often essentially simplify a spatial problem; on the other hand graphical methods of descriptive geometry can be applied more directly than those of analytic geometry. Further they are best suited to provide a check of computed numerical results–and a check of reasoning too. These problems are used at the Technical University Vienna to train engineering students. In descriptive geometry courses geometry is not only employed for drawing but it is also viewed to be primarily fundamental for understanding spatial relations.

Descriptive geometrical methods of space restitutions of algebraic plane curves

Abstract: A space curve is determined by intersection of surfaces. A plane curve is obtained by projection of a space curve. The aim of the space restitution of a plane curve is to determine whose curve of intersection is projected into that plane curve. The knowledge of these surfaces enables the descriptive geometrical treatment of a plane curve, that is, the construction of its points, tangents (in regular and singular points of the curve), asymptotes, focuses and the centres of osculatory circles in the vertices of the curve. The paper deals with two kinds of space restitution of plane curves: on the basis of their form and on the basis of their planimetric construction. The heuristic value of space restrictions is in the discovery of new forms of curves and surfaces.

12.5

Fundamental ideas for computer-supported descriptive geometry

Author: Paukowitsch, Peter
Corporate Source: Technische Univ, Vienna, Austria
Source: Computers & Graphics (Pergamon) v 12 n 1 1988 p 3-14
Publication Year: 1988
CODEN: COGRD2 ISSN: 0097-8493
Language: English
Document Type: JA; (Journal Article) Treatment: T; (Theoretical)
Journal Announcement: 8807

Abstract: Using the axonometric method for the uniform realization of parallel and central projection, different problems of descriptive geometry can be solved by means of a uniform algorithm, based on a simple geometric idea. Axonometry is introduced as a method for the uniform realization of parallel and central projection. An algorithm, based on a simple geometric idea, for the construction of a geometrically defined curve, which is contained in a surface, is demonstrated. This algorithm, using a uniform method, is able to solve different types of problems of descriptive geometry. The programs are written for the BASIC-pocket computer SHARP 1500A, and the figures are drawn by the plotter SHARP CE-150 and CE-516P, respectively. For most of the programs, only 5.9 kb of the 1500A are needed. (Edited author abstract) 4 refs.

12.6

Localized intersections computation for solid modelling with straight homogeneous generalized cylinders

Author: Ponce, Jean; Chelberg, David

Corporate Source: Stanford Univ, CA, USA

Sponsor: IEEE, Council on Robotics & Automation, New York, NY, USA; North Carolina State Univ, Raleigh, NC, USA

E.I. Conference No.: 10040

Source: Publ by IEEE, New York, NY, USA. Available from IEEE Service Cent (Cat n 87CH2413-3), Piscataway, NJ, USA p 1481-1486

Publication Year: 1987

ISBN: 0-8186-0787-4

Language: English

Document Type: PA; (Conference Paper)

Journal Announcement: 8709

Abstract: A report is presented on the development of a solid modeling system combining straight homogeneous generalized cylinders through set operations. Two basic components of this system are the modules which compute the set operations between primitives and display the resulting solids using ray tracing. These two modules are also very computationally intensive as they involve a large number of surface-surface and ray-surface intersections computations. The authors introduce a novel hierarchical representation for straight homogeneous cylinders called Box Tree. The Box Tree is analogous to a Quadtree in parameter space. It is an exact boundary representation which describes the surface of the associated generalized cylinder by a hierarchy of enclosing boxes. The Box Tree is used to efficiently compute the set operations and ray tracing algorithms by localizing the search for intersections to the regions where they may occur. A discussion is presented of complex issues and the performance of the modeling system on a variety of examples. 26 refs.

12.7

Graphic interaction in solids modelling

Molari, P.G.

Source: Pixel, Comput. Graph. CAD/CAM Image Process. (Italy) vol.9, no.3, pp.: 45-8

Publication Year: 1988

CODEN: PIXEDD ISSN: 0392-8217

Treatment: PRACTICAL;

Document Type: JOURNAL PAPER

Languages: Italian

(7 Refs)

Abstract: Solid modelling in mechanical design procedures is analysed underlining range of possibilities in augmenting traditional drafting. It is shown that solids modelling allows a three dimensional representation and dimensioning of an object or machine in a single entirety with perfect visualization of all lines and points. Opportunities offered by modelling based on 'constructive geometry' are enumerated showing pictorial examples of modelling.

12.8

Representation of straight lines by homogeneous 6-tuples and geometrical transformations

Barsanti, L.
Author Affil: Inst. di Matematiche Applicate, Univ. Degli Studi, Pisa, Italy
Source: Automatika (Yugoslavia) vol.29, no.1-2, pp.: 37-40
Publication Year: 1988
CODEN: ATKAAF ISSN: 0005-1144
Conference Information:
Third International Conference on Computer Graphics Date: 22-24 June 1988
Location: Dubrovnik, Yugoslavia
Treatment: THEORETICAL;
Document Type: CONFERENCE PAPER
Languages: English
(2 Refs)
Abstract: Deals with the geometrical transformations in the three-dimensional real projective space of a straight-line which is represented by a point in the five-dimensional real projective space. In descriptive geometry, this representation is useful to represent the transformations of the images of a straight-line on the three principal plane (H,F,P).

12.9

Fundamental ideas for computer-supported descriptive geometry

Paukowitsch, P.
Author Affil: Inst. fur Geometrie, Tech. Univ., Wien, Austria
Source: Comput. Graph. (UK) vol.12, no.1, pp.: 3-14
Publication Year: 1988
CODEN: COGRD2 ISSN: 0097-8493
U. S. Copyright Clearance Center Code: 0097-8493/88/$3.00 + .00
Treatment: THEORETICAL;
Document Type: JOURNAL PAPER
Languages: English
(4 Refs)
Abstract: Using the axonometric method for the uniform realization of parallel and central projection, very different problems of descriptive geometry can be solved by means of a uniform algorithm, based on a simple geometric idea.

12.10

The mathematical calculations involved in programming with LOGO to derive solutions to descriptive geometry problems

Ritter, R.L.
Author Affil: Div. of Fine Arts, Loyola Marymount Univ., Los Angeles, CA, USA
Sponsor: California Educ. Comput. Consortium

Source: 1986 Tenth Western Educational Computing Conference pp.: 90-6
Publication Year: 1986
Conference Information:
Date: 20-21 Nov. 1986 Location: Irvine, CA, USA
Publ: California Educ. Comput. Consortium, Sacramento, CA, USA. x+293 pp.
Treatment: PRACTICAL
Document Type: CONFERENCE PAPER
Languages: English
(4 Refs)

Abstract: The computer friendly nature of the LOGO language makes it an attractive tool for the very young who find the simple commands an easy and rather exciting method to use in developing a variety of geometric shapes and designs. This presentation addresses a more mature audience, the high school senior, and/or the college freshman, and in each case the teacher. Programming with LOGO to develop graphical solutions to problems in descriptive geometry is discussed.

12.11

Descriptive geometry and engineering drawing education using CADAM

Author: Giambalvo, Peter A.

Corporate Source: California Polytechnic State Univ, San Luis Obispo, CA, USA

Conference Title: Papers Presented at the 1986 Summer Meeting American Society of Agricultural Engineers.

Conference Location: San Luis Obispo, CA, USA Conference Date: 1986 Jun 29-Jul 2

Sponsor: ASAE, St. Joseph, MI, USA

E.I. Conference No.: 08900

Source: Paper - American Society of Agricultural Engineers Publ by ASAE, St. Joseph, MI, USA Pap 86-5025, 4p

Publication Year: 1986

CODEN: AAEPCZ ISSN: 0145-0166

Language: English

Document Type: PA; (Conference Paper)

Journal Announcement: 8702

Abstract: California Polytechnic State University has become a forerunner in the training of agricultural engineering students in the use of computer assisted design and computer assisted manufacturing. The training is met in courses on descriptive geometry and engineering drawing. Cal Poly's plan to incorporate CAD/CAM into the curriculum is fourfold: to involve a variety of career areas; to provide instruction from freshman to graduate levels; to expose students to various industrial approaches; and to apply CAD/CAM to existing curricula. This paper discusses the application. (Edited author abstract)

12.12

Combining descriptive and computational geometry

Author: Adams, J. Alan; Billow, Leon M.

Corporate Source: US Naval Acad, Annapolis, MD, USA

Source: Computers in Mechanical Engineering v 4 n 6 May 1986 p 69-76

Publication Year: 1986
CODEN: CMENDY ISSN: 0745-9726
Language: English
Document Type: JA; (Journal Article) Treatment: G; (General Review)
Journal Announcement: 8610

Abstract: Descriptive geometry is an established means for analyzing 3-D problems while working in a 2-D plane, such as a drafting table; and for transferring information from a 2-D plane, perhaps a blue print, back into the 3-D world. Orthographic projection and auxiliary views are the basis for communicating and interpreting this 3-D information in drawings or on computer displays. Computational geometry combines mathematical techniques classically associated with Euclidean, analytical, and differential geometry, as well as matrix algebra. Engineering students need the ability to visualize, reason, and communicate ideas in 3-D, and the ability to convert them into 2-D representations, and vice versa. This requires descriptive geometry techniques. On the other hand, learning CAD and CAM requires a previous understanding of computational geometry. In this article the authors compare both types of geometry and suggest that both objectives can be accomplished during undergraduate study by expanding descriptive geometry to include computational geometry, and truly taking advantage of the computer's ability. 10 refs.

12.13

Instructional modules in descriptive geometry

Author: Dix, Rollin C.; Howells, Arthur
Corporate Source: Illinois Inst of Technology, Chicago, IL, USA

Conference Title: Proceedings of the Sixth Annual Conference & Exposition - National Computer Graphics Association, Computer Graphics '85.
Conference Location: Dallas, TX, USA Conference Date: 1985 Apr 14-18
Sponsor: Natl Computer Graphics Assoc, Fairfax, VA, USA
E.I. Conference No.: 07045
Source: v 3. Publ by Natl Computer Graphics Assoc, Fairfax, VA, USA p 378-386
Publication Year: 1985
ISBN: 0-941514-09-9
Language: English
Document Type: PA; (Conference Paper)
Journal Announcement: 8510

Abstract: GRAFIX, the CAD introduction program in use at ITT, contains a set of commands which emulate the construction techniques of descriptive geometry. The paper describes a student exercise teaching the use of these commands, the graphics produced and the educational value of the experience. (Author abstract)

12.14

Descriptive geometry of the celestial sphere

Author: Brinker, Dexter M.

Conference Title: Technical Papers, 45th Annual Meeting - ACSM: Theodolite to Satellite. (Papers presented at the 1985 ASP-ACSM Convention (American Society of Photogrammetry).)

Conference Location: Washington, DC, USA Conference Date: 1985 Mar 10-15
Sponsor: American Congress on Surveying & Mapping, Falls Church, VA, USA
E.I. Conference No.: 06687
Source: Technical Papers of the American Congress of Surveying and Mapping 45th Annual Meeting. Publ by American Congress on Surveying & Mapping, Falls Church, VA, USA p 539-545
Publication Year: 1985
CODEN: TPAMDF ISSN: 0748-3244
Language: English
Document Type: PA; (Conference Paper)
Journal Announcement: 8508
Abstract: In past years, various authors have attempted to describe surveying astronomy and the PZS triangle. Many have fallen somewhat short of their goal because of the poor quality of their illustrations. Perhaps the best diagrams available to surveyors appeared in 'Surveying for Civil Engineers' by Philip Kissam, which was published in 1956. This paper discusses how to make a drawing of the celestial sphere which is easy to comprehend.

12.15

Geometric modeling of three-dimensional objects in automated planning systems

Pokhavlinskiy, A.B.; Visikirskiy, V.A.
Author Affil: Glushkov Inst. of Cybern., Acad. of Sci., Kiev, Ukrainian SSR
Source: Avtomatika (USSR) vol.18, no.6, pp.: 28-34
Publication Year: Nov.-Dec. 1985
CODEN: AVTMA8 ISSN: 0038-5328
Trans. in: Sov. J. Autom. & Inf. Sci. (USA) vol.18, no.6 pp. 20-3
Publication Year: Nov.-Dec. 1985
CODEN: SJISEL ISSN: 0882-570X
U. S. Copyright Clearance Center Code: 0882-570X/86/$7.50/0
Treatment: APPLIC; THEORETICAL;
Document Type: JOURNAL PAPER
Languages: English
(6 Refs)
Abstract: An examination is made of a combined model for description of the geometry of three-dimensional objects, including interconnected submodels of constructive geometry–a piecewise-analytic submodel and a wire-frame submodel. The advantages are considered of the use of each of the submodels at the stages of input, processing and output of information. Certain questions of simplification of the algorithms of set-theoretic operations over objects are investigated.

12.16

Computer graphics and teaching descriptive geometry

Waldron, M.B.; Hang, R.I.
Author Affil: Dept. of Eng. Graphics, Ohio State Univ., Columbus, OH, USA
Duncan, K.; Harris, D. (Editors)
Sponsor: IFIP

Source: Computers in Education. Proceedings of the IFIP TC3 4th World Conference on Computers in Education - WCCE 85 pp.: 657-61
Publication Year: 1985
Conference Information:
Date: 29 July-2 Aug. 1985 Location: Norfolk, VA, USA
Publ: North-Holland, Amsterdam, Netherlands. xxii + 1038 pp.
ISBN 0-444-87797-5
Treatment: PRACTICAL;
Document Type: CONFERENCE PAPER
Languages: English
(3 Refs)

Abstract: A software algorithm to teach descriptive geometry in the classroom is presented. The method of rotation is revisited. The ease of rotation of objects using computers makes this method very powerful and eliminates messy construction lines necessary in conventional methods of teaching descriptive geometry. The philosophy in developing the software is to allow the student to visualize the object as it rotates in space rather than getting lost in the mechanics of constructional detail. The design is modular and problem entry is easy. The high resolution of the Tektronix screen and the computational capabilities of the IBM 3277GA hosted by a 4341 computer is utilized to its fullest.

12.17

Instructional modules in descriptive geometry

Dix, R.C.; Howells, A.
Author Affil: Illinois Inst. of Technol., Chicago, IL, USA

Sponsor: Nat. Comput. Graphics Assoc.
Source: Proceedings of the Sixth Annual Conference and Exposition: Computer Graphics '85. Vol. III, Technical Sessions pp.: 378-86
Publication Year: 1985
Conference Information:
Date: 14-18 April 1985 Location: Dallas, TX, USA
Publ: Nat. Comput. Graphics Assoc., Fairfax, VA, USA. 871 pp.
ISBN 0-941514-09-9
Treatment: APPLIC;
Document Type: CONFERENCE PAPER
Languages: English

Abstract: GRAFIX, the CAD introduction program in use at IIT, contains a set of commands which emulate the construction techniques of descriptive geometry. The paper describes a student exercise, teaching the use of these commands, the graphics produced and the educational value of the experience.

12.18

The application of descriptive geometry in determination of the radiation angle factor

Ying Yufang; Yu Qizheng

Source: J. Harbin Inst. Technol. (China) no.1, pp.: 126-9

Publication Year: 1985
CODEN: HPKYAY
Treatment: THEORETICAL;
Document Type: JOURNAL PAPER
Languages: Chinese
(2 Refs)
Abstract: The authors introduces a graphic method to calculate the radiation angle factor by determining the angles. This is the application of descriptive geometry in determining the angle factor of radiation heart transfer. For some figures this method is simpler and the accuracy is in the range of engineering application.

12.19

Computer aided instruction of descriptive geometry
Smith, L.F.; Wolf, J.J., III; Hopkins, M.R.; Jou, D.T.
Author Affil: Ohio State Univ., Columbus, OH, USA
Source: 26th ADCIS Conference Proceedings pp.: 169-74
Publication Year: 1985
Conference Information:
Date: 25-29 March 1985 Location: Philadelphia, PA, USA
Publ: ADCIS, Bellingham, WA, USA. 273 pp.
Treatment: PRACTICAL;
Document Type: CONFERENCE PAPER
Languages: English
Abstract: Descriptive Geometry, that segment of graphic science which determines the relationship between points, lines and surfaces, traditionally has been practiced with T-squares, triangles, protractors, compass etc. An interactive computer program, DEGE, was written to replace the traditional 'tools' with a two screen interactive workstation. The interactive workstation, assembled and used by the Interactive Graphics Laboratory of the Ohio State University, contains the following components: IBM 3277 Display Station; Tektronix Graphics Monitor-618; Cursor Control Unit; Tektronix 4611 Hard Copy Unit; and an IBM 4341 Computer. The effectiveness of DEGE is discussed.

12.20

Application of computer graphics for descriptive geometry
Ezaki, T.; Jensen, J.K.
Source: Fukuoka Univ. Rev. Technol. Sci. (Japan) no.34, pp.: 91-4
Publication Year: March 1985
CODEN: FDKSDF ISSN: 0285-2799
Treatment: APPLIC;
Document Type: JOURNAL PAPER
Languages: Japanese
(3 Refs)
Abstract: Recently, because of the availability of high performance personal computers which have graphics functions there has been an interest in using the computer as an extension of instrument drawing and effective educational facilities. The authors discuss the use of computer graphics for descriptive geometry and a

method of data input. The described method utilizes graphic software in a large class setting.

12.21

Algorithm for determining the intersection of two simple polyhedra
Author: Szilvasi-Nagy, M.
Corporate Source: Technical Univ of Budapest, Dep of Geometry, Budapest, Hung
Source: Comput Graphics Forum v 3 n 3 Sep 1984 p 219-225
Publication Year: 1984
CODEN: CGFODY ISBN: 0-306-41436-8
Language: English
Document Type: JA; (Journal Article) Treatment: T; (Theoretical)
Journal Announcement: 8503
Abstract: An algorithm is presented for the construction of the intersection of two simple possibly non-convex polyhedra. The methods of descriptive geometry are applied, so that this three-dimensional problem can be solved in two dimensions. All intersections of the edges of each polyhedron with the faces of the other are sought, and these points of intersection can be found in Monge's model. Projecting both polyhedra on the xy and on the xz coordinate planes we obtain two superimposed maps in both projections. The algorithm to find the intersection of two maps is based upon the Shamos-Hoey, the Bentley-Ottmann and the Nievergelt-Preparata algorithms. The asymptotic time requirement for determining the polygon of intersection of two polyhedra is 0((N PLUS S) log N), where N is the sum of the numbers of vertices of the two polyhedra and S is the total number of intersections of all projected edges in the xy-plane (S = 0(N**2)). 4 refs.

12.22

On the condition of equality between a spatial angle consisting of two oblique lines and its projective angle
Lu Xinlong
Author Affil: Shanghai Chiao Tung Univ., Shanghai, China
Source: Shanghai Jiaotong Daxue Xuebao (China) No. 4, pp.: 165-76
Publication Year: 1983
CODEN: SCTPDH
Treatment: APPLIC; THEORETICAL;
Document Type: JOURNAL PAPER
Languages: Chinese
(7 Refs)
Abstract: In descriptive geometry, the relations between some specific special angles and their protective angles are generally described through a sort of intuitive geometrical approach. So far, however, no proof has yet been made of the relation between a spatial angle consisting of two oblique lines and its projective angle. A general rule for determining whether a spatial angle consisting of two oblique lines is equal to, greater or less than its projective angle has been obtained by applying the analytical method. A facile construction method of equalizing a projective angle to a spatial angle in orthographic projection is presented. Finally, on the basis of the

principles presented, the conditions and the calculation method for developing the convex surfaces of a regular convex polygon to a complete star figure are given. The shortcomings of the conventional construction method for developing convex surfaces from the convex five-pointed star are pointed out, and a brand-new facile construction method is introduced.

12.23

Computer aided design and descriptive geometry for school children
Edmonds, G.
Author Affil: Fortismere School, London, England
Source: Comput. Educ. (GB) No. 46, pp.: 2-4
Publication Year: Feb. 1984
CODEN: CPECBK ISSN: 0010-4590
Treatment: PRACTICAL;
Document Type: JOURNAL PAPER
Languages: English
(2 Refs)
Abstract: Descriptive geometry is an exact science that deals with the basic theory of representation. The use of descriptive geometry occurs in various forms throughout the disciplines of engineering, science, and architecture. The representation of design concepts through drawings requires an understanding of drawing that most young pupils do not have. PICASO—picture computer algorithms subroutine orientated—is a computer language designed to enable graphics peripherals to be used with a minimum of programming effort and skill. Although the system was initially designed for artists and designers, it includes extensive facilities that are relevant to mathematics and engineering. PICASO consists entirely of FORTRAN subroutines and functions. These are normally held in a precompiled form as a library, and are referenced by FORTRAN programs or any other language permitted to operate in this mode.

12.24

Constructive geometry in three dimensions for computer-aided design
Author: Sharpe, R. J.; Thomas, P. J.; Thorne, R. W.
Corporate Source: Summagraphics Corp, Fairfield, Conn, USA
Source: Journal of Mechanical Design, Transactions of the ASME v 104 n 4 Oct 1982 p 813-816
Publication Year: 1982
CODEN: JMEDDD ISSN: 0161-8458
Language: ENGLISH
Journal Announcement: 8301
Abstract: Three dimensional construction techniques are described along with the detailed description of several derivations. The resultant algorithms allow construction of points, lines, and conics directly in 3-space. These pieces of geometry are constructed relative to existing lines, conics, and splines which are represented as parametric rational cubics. The generality of these algorithms allows a single routine to handle construction relative to all curve types in the system. This characteristic reduces the amount of code and thus the implementation and maintenance costs of

the CAD system. These techniques are being incorporated into a Computer-Aided Design system. 7 refs.

12.25

On an educational method for computer graphics introduced into descriptive geometry

Nagae, S.; Tsujimura, K.; Fukunaga, S.
Author Affil: Course of Math. and Related Fields, Coll. of Integrated Arts and Sci., Univ. of Osaka, Osaka, Japan

Source: Bull. Univ. Osaka Prefect. Ser. A (Japan) vol. 29, no.1, pp.: 73-81
Publication Year: 1980
CODEN: BSKAAJ
Treatment: APPLIC; PRACTICAL;
Document Type: JOURNAL PAPER
Languages: English
(4 Refs)

Abstract: Reports on an educational method for teaching computer graphics, which has been given as a part of a first year descriptive geometry course for engineering students. In general it can be said that the effective teaching of computer graphics requires the ability of instructors to instil in students an understanding of the characteristics and actions of given figures in relation to the theory of descriptive geometry.

12.26

Set membership classification: a unified approach to geometric intersection problems

Tilove, R.B.
Author Affil: Coll. of Engng. and Appl. Sci., Univ. of Rochester, Rochester, NY, USA

Source: IEEE Trans. Comput. (USA) vol.C-29, no.10, pp.: 874-83
Publication Year: Oct. 1980
CODEN: ITCOB4
Treatment: THEORETICAL;
Document Type: JOURNAL PAPER
Languages: English
(18 Refs)

Abstract: Many practically important problems in computational geometry may be regarded as a generalization of clipping, and may be formulated precisely in terms of a function called membership classification. This function operates on a pair of point sets called the reference and candidate sets; it segments the candidate into three subsets which are 'inside,' 'outside,' and 'on the boundary of' the reference. Examples of classification problems include clipping, polygon intersection, point inclusion, and solid interference. The paper has two major parts. In the first, classification is defined abstractly in terms of regular sets and regularized set operators. The second part is devoted to a study of algorithms for classifying candidates with respect to reference sets represented constructively, i.e., as combinations (via the regularized set operators) of primitive regular sets. The author presents algorithms that are both general and efficient, and that are used in a major geometric processor developed at the University of Rochester.

12.27

Descriptive geometry—it can provide the solution to some design problems

Author: Farrar, J. A.
Source: Chartered Mechanical Engineer v 26 n 10 Nov 1979 p 39-42
Publication Year: 1979
CODEN: CHMGAF ISSN: 0009-191X
Language: English
Journal Announcement: 8005

Abstract: The article outlines some fundamentals applied in descriptive geometry to facilitate and clarify the projection of ancillary views of a more complex constructional nature. The designer's time is spent predominantly in producing drawings in which there is marginal need for the application of a datum in the manner outlined. Consequently when a design layout drawing of more than usual complexity is under construction neglect of these fundamentals becomes all too apparent. Application of the principle of using ground lines, with the identification of major points, on every design layout drawing requiring more than three principal views would establish a discipline of merit for future unforeseen eventualities.

12.28

A microprocessor system to help teach descriptive geometry

Zsombor-Murray, P.J.
Author Affil: Dept. of Mech. Engng., McGill Univ., Montreal, Canada
Hamza, M.H. (Editors)
Sponsor: IEEE
Source: Proceedings of the International Symposium on Mini and Micro Computers pp.: 172-8
Publication Year: 1978
Conference Information:
Date: 11-18 Nov. 1977 Location: Montreal, Canada
Publ: IEEE, New York, USA. VIII + 318 pp.
Treatment: APPLIC; PRACTICAL;
Document Type: CONFERENCE PAPER
Languages: English
(5 Refs)

Abstract: A computer aided instruction language, used to help teach descriptive geometry and consisting of a 3-word vocabulary with modifying suffixes, is described. A sample exercise, performed by the current implementation on a medium scale computer with an online plotter, is presented. The implementation of this teaching aid on a microprocessor with an inexpensive plotter provides, apart from the rote translation of the existing software, an interesting problem in 2-axis, straight line motion control.

12.29

Suggestions for the application of a plotter in descriptive geometry
Seybold, H.
Author Affil: Inst. fur Math., Tech. Univ. Munchen, Munchen, Germany
Source: Computing (Austria) vol. 17, no.4, pp.: 281-7
Publication Year: 1977
CODEN: CMPTA2
Treatment: APPLIC; THEORETICAL;
Document Type: JOURNAL PAPER
Languages: German
Abstract: Simple methods for solving problems of nonlinear descriptive geometry in R/SUP 3/ by means of a computer and a plotter are suggested: Automatic contructions of contours of surfaces in parallel and central projection and of parallel and central projections of curves of intersection of two surfaces with hidden lines omitted or dotted. The methods are generally applicable to all surfaces which are given by an equation and by equations in parametric form.

12.30

Computer exercises in engineering graphics
Author: Zsombor-Murray, Paul
Corporate Source: McGill Univ, Montreal, Que
Source: COED Transactions (Computers in Education Division of ASEE) v 9 n 12 Dec 1977 p 153-189
Publication Year: 1977
CODEN: ASCTCR
Language: English
Journal Announcement: 7806
Abstract: The role of graphics in engineering curricula and its similarity to the role of programming is discussed. A survey on computer graphics in education, including specific, relevant literature, is included. The development of a one semester engineering graphics course, with emphasis on a computer graphics module contained therein, is described. Examples of exercises and tests are presented. Topics include labeling, graph plotting, digitization of graphical data, a descriptive geometry interpreter, perspective view generation, two contour mapping methods and a representative mechanism animation assignment.

Index

R

S

T

U

V

W